Equine-Assisted Therapy and Learning
with At-Risk Young People

Equine-Assisted Therapy and Learning with At-Risk Young People

Hannah Louise Burgon
Sirona Therapeutic Horsemanship CIC, UK

© Hannah Louise Burgon 2014
Foreword © Leif Hallberg 2014

All rights reserved. No reproduction, copy or transmission of this publication may be made without written permission.

No portion of this publication may be reproduced, copied or transmitted save with written permission or in accordance with the provisions of the Copyright, Designs and Patents Act 1988, or under the terms of any licence permitting limited copying issued by the Copyright Licensing Agency, Saffron House, 6–10 Kirby Street, London EC1N 8TS.

Any person who does any unauthorized act in relation to this publication may be liable to criminal prosecution and civil claims for damages.

The author has asserted her right to be identified as the author of this work in accordance with the Copyright, Designs and Patents Act 1988.

First published 2014 by
PALGRAVE MACMILLAN

Palgrave Macmillan in the UK is an imprint of Macmillan Publishers Limited, registered in England, company number 785998, of Houndmills, Basingstoke, Hampshire RG21 6XS.

Palgrave Macmillan in the US is a division of St Martin's Press LLC, 175 Fifth Avenue, New York, NY 10010.

Palgrave Macmillan is the global academic imprint of the above companies and has companies and representatives throughout the world.

Palgrave® and Macmillan® are registered trademarks in the United States, the United Kingdom, Europe and other countries.

ISBN 978–1–137–32086–5

This book is printed on paper suitable for recycling and made from fully managed and sustained forest sources. Logging, pulping and manufacturing processes are expected to conform to the environmental regulations of the country of origin.

A catalogue record for this book is available from the British Library.

A catalog record for this book is available from the Library of Congress.

Transferred to Digital Printing in 2015

Contents

List of Illustrations	vii
Foreword	viii
Preface	x
Acknowledgements	xi
List of Abbreviations	xiii

Introduction	1
1 Background to Equine-Assisted Therapy and Learning	6
2 Young People Considered 'At Risk', and Overview of 'The Yard' and Methods Used	35
3 Development of Self-Confidence and Self-Efficacy, and the Opening Up of 'Positive Experiences' and 'Positive Opportunities' through Therapeutic Horsemanship	83
4 Developing Attachments, Empathy and Trust through Relationships with Horses	105
5 The Horse, the Therapeutic Relationship and Other Psychotherapeutic Insights	125
6 Horses, Mindfulness and the Natural Environment	154
7 Conclusion	190

Appendix A: Research Project into Equine-Assisted Therapy/Learning and Therapeutic Horsemanship	207
Appendix B: Consent Form – Young People	209
Appendix C: Research Project into Equine-Assisted Therapy/Learning and Therapeutic Horsemanship	211
Appendix D: Consent Form – Adults	212
Appendix E: XXXX Ltd Therapeutic Horsemanship Programme	213

vi *Contents*

Appendix F: XXXX Ltd Therapeutic Horsemanship Programme 215

*Appendix G: Questionnaire about XXX Therapeutic Horsemanship
(children and young people)* 216

Appendix H: XXX Therapeutic Horsemanship Programme 217

Appendix I: PhD Questions for Young People 218

Appendix J: Tables of Participants 220

Glossary 222

Bibliography 223

Index 247

Illustrations

All photographs are by Shaun Dérioz and posed by models.

1 Making friends: having a scratch
2 Horse care: plaiting a mane
3 Horse care: picking out hooves
4 Horse care: grooming
5 Shetlands like hugs too
6 Group 'invisible leading' session
7 Exercises in the round pen: walking in rhythm
8 Exercises in the round pen: follow the leader
9 Exercises in the round pen: join-up
10 Horse agility: negotiating obstacles
11 Horse agility: building trust and confidence
12 Horse agility: crossing bridges
13 Horse agility: investigating the box
14 Horse agility: testing the box
15 Horse agility: success!
16 Exercises on horseback: stretching and balance
17 Exercises on horseback: balance and confidence
18 Equine-assisted education session
19 Equine-assisted education session: Jasper the horse helping with worksheet
20 Taking some time out together
21 Good friends
22 Getting to know each other: hand grooming

Foreword

At 37 years old it's amazing to be able to say that I remember the birth of a field, but I can and I do. I was in undergraduate college at the time, a horsewoman studying environmental journalism, and never knew that what I was about to witness would change my life forever.

In the 1990s a series of events transpired in the USA that would shift the paradigm of the horse–human relationship. Starting with Barbara Rector's creation of an equine-facilitated psychotherapy programme at Sierra Tucson in Tucson, Arizona, and continuing to include the subsequent development of national and international membership organisations, colleges and universities providing degrees and coursework, and thousands of professionals offering services, the movement to include equines as a part of a treatment or learning team was officially born.

I watched as the field exploded, crossing coast to coast and continent to continent. Around the globe, people began viewing horses and the horse–human relationship differently. For many, this orientation comes without the need for explanation or validation. It comes from a heartfelt sense of connection and belief. For others, research and rigorous academic study hold the key to the field's long-term success.

My role within this field has been not only to provide services, delving deeply into the felt experience of the work, but probably more importantly to act as an observer and scribe, documenting the development of the profession.

It is because of this that I first came to encounter Dr Hannah Louise Burgon. I was writing my first book when I came across her research. In 2003, Anthorozoos published the results of a study that she conducted while at the University of Exeter in the UK, entitled 'Case Studies of Adults Receiving Horse-Riding Therapy'. The purpose of this study was to examine the psychotherapeutic results of therapeutic riding. By the time the article was published, a number of studies had already been conducted which investigated the physiological effects of horse-riding, but little had been done to understand the possible psychological benefits. Dr Burgon's work was cutting edge and her contribution helped to move the field forward significantly.

In reading the results of her first study, I noticed immediately that Dr Burgon was not afraid to ask tough questions. Her enquiry into the after-effects of the treatment method was a topic that many had veered away from in both practitioner conversation and prior studies. She demonstrated that although passionate and invested, she was not going to allow her own opinions and desires to cloud her professionalism and dedication to 'real' science.

This characteristic is now seen in her first book, *Equine-Assisted Therapy and Learning with At-Risk Young People*. In this book, Dr Burgon has interwoven exhaustive and scholarly research with personal experience and powerful vignettes. In this way she has been able to speak as both a researcher and a provider, giving us a unique perspective on the 'how' of the work. *Equine-Assisted Therapy and Learning with At-Risk Young People* also includes both historic and current considerations, emerging trends and theories, and it builds a strong foundation from which to present the research results. Dr Burgon's work introduces an international perspective, and continues to build the conceptual framework from which further studies can be developed.

This book marks a significant contribution to our body of knowledge. Dr Burgon has effectively engaged our hearts and our minds, validating our personal experiences while providing us with language to explain our feelings. I am in awe of how far our field has come and am honoured to share it with such knowledgeable, thoughtful and skilled professionals as Dr Burgon.

Leif Hallberg, MA, LCP
Author/Psychotherapist/Consultant
Portland, Oregon, USA

Preface

This book explores the experiences of seven 'at-risk' young people who participated in a unique therapeutic horsemanship programme based in the UK where the research study took place. Therapeutic horsemanship is aligned to the developing fields of equine-assisted psychotherapy and equine-assisted learning, where horses are utilised alongside experienced practitioners for social, emotional and learning benefits.

The young people who attended this programme were referred from various organisations, including a foster care agency, youth offending team and a pupil referral unit, and were considered to be 'at risk' due to their various psychosocial disadvantages. In addition to the young people, the views of some of the adults involved with them were included. These were therapists on the therapeutic horsemanship programme, foster carers, a teacher and a youth offending team mentor.

A number of prominent themes were uncovered from the research study. These were loosely based around issues of nurture, attachment and trust; social well-being and resilience; identification with the horse; a 'safe' space and calming influence; the role of the horse in the therapeutic arena; and the natural environment and spiritual dimensions. Relevant theoretical frameworks adopted were from the risk and resilience literature; attachment theory; psychotherapeutic literature; mindfulness; and the nature/outdoor and ecotherapy fields.

Acknowledgements

First and foremost, I wish to express my deep appreciation to the young people who so generously and graciously participated in the research study on which this book is based. The openness that they displayed in sharing their experiences and thoughts with me was something that I feel privileged to have been part of. I am also hugely grateful to the adults who gave up their time to be interviewed and provided me with so much additional food for thought.

So many others helped to make this project possible. Without my husband Shaun's unwavering support, it would never have happened: thank you a million times for everything, and also for the amazing photographs you take. I am grateful to all of my family for their belief in me and for being there. Thanks to Di Gammage for our many conversations about young people and horses and much more besides. I am forever indebted to Lucy Rees who started all of this by sitting me on a half-wild Welsh mountain pony aged seven or eight, and encouraging and supporting me with horses and numerous other things ever since. I hope that I can share with many other young people just some of the experiences that Lucy shared with me.

I would like to acknowledge some of the pioneers in this exciting new field of equine-assisted therapy and learning: Barbara Rector, whose training clinic in the UK I had the good fortune to attend during my early foraging in this world; and also Linda Kohanov and Adele and Marlena McCormick, whose books were the first that I read and that inspired me to want to find out more. It was then exciting to discover that Leif Hallberg, who has generously provided a foreword to this book, had also written in her important book, *Walking the Way of the Horse*, about themes and concepts similar to those that I uncovered. This hopefully helps to provide further credibility and credence to the field of equine-assisted therapy and learning – the fact that researchers and practitioners from across the world are discovering and documenting the same themes and benefits that people can gain from contact with horses.

I dedicate this book to my father, who so sadly died before I completed it. His unwavering determination and dedication to follow his inner passion and calling to music despite the challenges that he encountered provided a role model that I did not always appreciate.

Lastly, but in no way least, I thank the horses, especially my beautiful Donzela (the fairy horse according to many of the young people fortunate to have known her), who continue to teach me every day.

Abbreviations

AAT	animal-assisted therapy
ADD	attention deficit disorder
ADHD	attention deficit hyperactivity disorder
CBEIP	Certification Board for Equine Interaction Professional
CBT	cognitive-behavioural therapy
DLA	disability living allowance
EAA	equine-assisted activities
EAC	equine-assisted counselling
EAGALA	Equine-Assisted Growth and Learning Association
EAI	equine-assisted interventions
EAL	equine-assisted learning
EAT/L	equine-assisted therapy and learning
EAP	equine-assisted psychotherapy
EAT	equine-assisted therapy
EFEL	equine-facilitated experiential learning
EFL	equine-facilitated learning
EFMHA	Equine Facilitated Mental Health Association
EFP	equine-facilitated psychotherapy
EGEA	Equine Guided Education Association
EI	temperament-based index of emotionality
GAF	Children's Global Assessment of Functioning Scale
LA	local authority
MBCT	mindfulness-based cognitive therapy
MBSR	mindfulness-based stress reduction
NGC	Natural Growth Centre
PAT	Pets as Therapy
PATH Intl.	Professional Association of Therapeutic Horsemanship International
PRU	pupil referral unit
RDA	Riding for the Disabled
RSW	registered social worker

xiv   *List of Abbreviations*

SCAS	Society for Companion Animal Studies
SEN	special educational needs
SW	social worker
TH	therapeutic horsemanship
WMP	Wild Mustang Program
YOT	Youth Offending Team

Introduction

During the last decade a novel way of working with people experiencing a range of emotional and social difficulties has started to gain recognition in the UK by the mental health and allied helping professions. This exciting new field utilises horses alongside trained professionals and is known variously as equine-assisted psychotherapy (EAP), equine-assisted therapy (EAT), equine-assisted learning (EAL), equine-assisted activities (EAA), equine-assisted counselling (EAC) and equine-facilitated mental health (EFMH), amongst others. There are also variations on these terms, utilising the word 'facilitated' instead of 'assisted' to different effect (Halberg, 2008). Whilst conversely Riding for the Disabled (RDA), where people with physical disabilities ride horses, was started in the UK, the field of EAT and EAL has a much longer established base in the USA where Rupert Isaacson's bestselling book *The Horse Boy*, about the experiences of his autistic son being helped by horses, has recently helped to give the field international publicity.

This book is about the experiences of seven young people who participated in a therapeutic horsemanship (TH) programme that I established whilst working as a social worker for a foster care agency in the south of England. It is part of a long journey that evolved from horses being a childhood passion and led to working with horses together with young people with various disadvantages and difficulties in a therapeutic and educational capacity. The research study outlined in this book followed as an attempt to articulate and share some of the experiences of 'The Yard', the TH programme on which this book is based.

Through a personal belief and understanding that contact with horses could be therapeutic in many ways, I found myself in the fortunate position to be able to incorporate TH in my work as a social worker for a foster care company. Whilst initially this was limited to young people who were in foster placements with this particular company, the TH programme expanded through interest from outside agencies to include

2 Equine-Assisted Therapy and Learning with At-Risk Young People

young people referred from youth offending and other organisations. The young people who attended the TH programme, named 'The Yard' for the purpose of this book, were generally considered to be 'at risk' within the social work and allied professions due to their circumstances and histories. Substantial research within the literature concerned with risk and resilience claims that there is a high likelihood of negative life outcomes as a result of experiencing adverse childhood experiences (Masten, Best & Garmezy, 1990; Rutter, 1985).

Definition of terms

Therapeutic horsemanship (TH) is one of the many different approaches within the fields of equine-assisted learning (EAL), equine-assisted therapy (EAT) and equine-assisted psychotherapy (EAP). For the sake of this book I employ the abbreviations EAL and EAT, which I have amalgamated into EAT/L, as they provide a broad description of the interventions and are the most commonly employed in the UK at the time of writing. The fields of EAT/L and TH have evolved alongside animal-assisted therapy (AAT) as additional therapeutic interventions with both young people and adults benefiting from therapeutic and/or learning support. They work from the premise that contact with animals is therapeutic and healing in a variety of different ways and can be useful for people who may find traditional therapeutic or educational interventions challenging or difficult in some way.

In a similar way to any emerging field, there are diverse differences in the use of terminology and little standardisation of practice. A number of organisations are attempting to provide frameworks and training to address some of these issues and create professionalism in the field. So far the most prominent have been established in the USA. These are the Professional Association of Therapeutic Horsemanship International (PATH Intl.) and the Equine-Assisted Growth and Learning Association (EAGALA), which has a European arm. Compared to a few years ago, an internet search will now provide a huge array of other organisations offering training in the respective fields – far too many to list here. They include the Certification Board for Equine Interaction Professional (CBEIP), the Equine Guided Education Association (EGEA), which offers training predominantly in equine-guided leadership and an annual conference, and Epona, the author Linda Kohanov's organisation, which has a number of trained practitioners now based in the UK. The *Federation of Horses in Education and Therapy International (HETI)* is a membership organisation and publishes an annual journal. Whilst there remain various differences between the organisations in terms of ideas about best practice, a general agreement appears to be that in order to employ the terminology equine-assisted or equine-facilitated psychotherapy (EAP/EFP), one member of the team is required to be a trained and registered mental health professional, such as a psychotherapist, psychologist or counsellor. Another general, although not universal, agreement is that two practitioners are required to facilitate an EAT/L or TH session in order to provide a safe environment for clients and horses. The Yard followed this guideline and, in addition to this, followed the principle that both practitioners needed to have adequate equine experience, and particular knowledge of the horses employed in the practice. Most EAP and many EAT/L practices do not offer riding. The emphasis instead is on encouraging participants to uncover insights and self-awareness through ground-based exercises and experiences with horses, together with the therapist providing intervention along their own particular style of practice. Often EAP and EAT/L are time-limited. In contrast, TH is often a more encompassing intervention which can include horse riding together with all aspects of horse care and management; it is understood to be therapeutic as opposed to being a therapy as such. Additionally, as The Yard employed a qualified play and drama psychotherapist, together with a counsellor, EAP/C was also provided.

Application of the book

This book explores why horses seem to appeal to and motivate young people who have suffered exclusion and disadvantage and who may have previously found it difficult to engage with services. It seeks to examine some of the factors about the horse that lends it to providing a unique therapeutic medium, its cultural role in history and links to biophilia – the study of the longstanding, historical and, it is claimed, innate relationship between people, nature and animals (Wilson, 1984). Due to the small sample size the findings in this book are not presented as having universal application. Instead I am interested in giving the reader a real sense of the experiences and views of a particular group of young people and associated adults who were involved in a unique TH programme in a particular location at a particular moment in time. However, located alongside other allied research in the field of EAT/L, themes arose that have many similarities and parallels and so it is hoped that this book can contribute in a small way to the growing evidence base behind this important new way of working with people and horses to improve lives.

4   *Equine-Assisted Therapy and Learning with At-Risk Young People*

When I initially established The Yard, I found a lack of research and literature around EAT/L available, and this largely remains the case. One hope from this book is that it will prove useful for other practitioners seeking to establish TH and EAT/L by presenting through the participants' own words and perceptions how they experienced TH. In addition, readers are able to gain an account of how other adults and practitioners involved in The Yard believed TH to be beneficial to the young people who were participating.

As there is very little research in the field of EAT/L and TH, it is hoped that this book will also contribute to the research base in the widest sense. Since the field is relatively new, a proportion of the available research is 'grey' literature in the form of PhD theses, dissertations and case-study accounts by practitioners and therapists. The academic literature that exists consists of a handful of studies utilising both qualitative and quantitative methodologies, but with limitations such as small populations, lack of control groups and short timeframes (Bachi, 2012; Smith-Osborne & Selby, 2010). In addition, it has been suggested that the current research emphasis, largely led by clinical psychology, has sought to focus on quantifying the effects of horse–human relations, and that this misses the intricacies of the relationships and experience, with it being argued that 'The *how* of what happens is equally important' (Garcia, 2010: 88).

This book seeks to provide an additional contribution to the research base by providing an in-depth account of the experiences of a number of young people participating in TH through a practitioner-researcher ethnography over a two-year period.

Overview of the book

In Chapter 1 the literature around EAT/L and the position of TH within the field is introduced, followed by a brief outline of the aligned field of AAT, which has a longer and more established research base. This followed the publication of a number of key studies in the 1980s that claimed that stroking a dog could lower blood pressure and that heart-attack victims who owned pets had greater life expectancy (Friedmann *et al.*, 1980, 1983). Despite the lack of research literature, it has been argued that the employment of horses in the therapeutic domain has a long history and there is evidence that Greek soldiers were encouraged to ride horses to aid recovery (All, Loving & Crane, 1999; Barclay, 1980). With the centrality of horses in civilisation throughout history and across cultures, they have held a unique place in human history. As a result, horses have continued to be deeply implicated in culture as symbolic icon through fairy tales, symbolism, film, art and poetry (Scanlan, 2001; Walker, 2008).

The chapter also looks at the place of animals in the sociological literature, briefly exploring children's position in this and the ways in which animals are employed to teach cultural norms and values. It also introduces themes centred on the natural environment and touches upon the literature that explores different aspects of nature and its importance in terms of human health, in particular for children and young people.

Chapter 2 introduces how the young people who participated in the study are positioned within the literature that is concerned with risk and resilience. Some of the theories behind the different understandings of risk and resilience and why some young people appear to overcome adversity despite negative childhood experiences are explored, together with what have been termed 'protective factors', and how some of these may have links to benefits from participation in EAT/L and TH. The chapter goes on to briefly introduce attachment theory and the role of the horse in the therapeutic relationship from related psychotherapeutic perspectives. An area that has so far received little attention in the EAT/L literature, that of the possible links between mindfulness and EAT/L and TH, is then explored. Finally, the chapter gives an overview of The Yard and a short description of the research methods employed in the research study.

The following chapters explore the themes that arose from the research. These are presented as linked to the theoretical frameworks in which they are positioned. Chapter 3 looks at the area of self-confidence, self-esteem and self-efficacy, and other themes linked to the risk and resilience literature. In Chapter 4 the subject of trust and empathy between the young people and horses, framed within attachment theories, is explored. Chapter 5 goes on to explore themes that arose from the psychotherapeutic literature as related to the role of the horse in the therapeutic relationship, together with those of identification and metaphor. Chapter 6 introduces links between TH and mindfulness, such as how being with horses can bring participants into the present moment and induce a feeling of calmness and well-being. In addition, the literature surrounding the natural environment and nature therapy is discussed and explored in relation to the psychosocial, psychospiritual and physical dimensions of horses and nature.

Chapter 7 concludes with some of the limitations and implications for policy and practice, together with suggestions for future research.

1
Background to Equine-Assisted Therapy and Learning

TH and EAT/L are interventions where therapists, social workers, teachers and other related professionals employ horses for therapeutic and/or learning benefit with young people and adults with a range of psychosocial issues. Improvements are reported in mental health and physical health, in addition to behavioural change and educational and learning opportunities. Alongside EAT/L, horses are starting to be utilised in the self-development and business leadership field in the UK. Here the emphasis is on personal growth and team-building for the purpose of achieving more effective skills and success in the corporate environment (e.g. see www.egld.eu and www.leadchange.com).

Whilst it has been argued that horses have been employed as healers for centuries (Broersma, 2007; Kohanov, 2001, 2005; McCormick & McCormick, 1997, 2004), EAT/L as a contemporary profession is still a relatively newly emerging field which has largely been led by the USA. Consequently, published studies on EAT/L are limited, until recently mainly consisting of descriptive and anecdotal accounts of these interventions by practitioners and therapists. There is a slightly larger research base on hippotherapy – the employment of the horse for physical therapy – which has a longer history, having been employed by occupational therapists and physiotherapists since the 1950s (Bertoti, 1988; Davies, 1988; Engel, 1984, 1997; Henrickson, 1971; Kunzle, Steinlin & Yasikoff, 1994; MacKinnon *et al.*, 1995; Young & Bracher, 2005; Young, 2005).

It is interesting that the emerging research on EAT/L largely follows the traditional approach to treating psychological problems, namely the medical model, with a large high proportion of recent studies utilising quantitative methods and psychological measuring techniques, and being published in psychology and health-based as opposed to sociologically orientated journals (Bachi, Terkel & Teichman, 2012; Bizub, Joy & Davidson, 2003; Bowers & MacDonald, 2001; Ewing, *et al.*, 2007; Rothe *et al.*, 2005; Schultz, Remick-Barlow & Robbins, 2007; Trotter *et al.*, 2008; Vidrine, Owen-Smith & Faulkner, 2002). A number of the limited studies on EAT/L published in sociological journals have also employed psychological testing techniques (Kaiser *et al.*, 2004; Klontz *et al.*, 2007). Surprisingly, considering the infancy of research in the field as a whole, a relatively large body of the available research in EAT/L involves young people. This is in contrast with AAT, where the majority of research was initially with adults. This may be due to the fact that EAT/L has evolved largely from the RDA field, which caters mainly for children and young people, whereas AAT was initially largely concerned with animals in the context of the health of adults and older people, such as guide dogs for the blind. However, as outlined above, the emphasis has been on research following a pre- and post-test design with psychological testing methods concerned with measuring change (Garcia, 2010). Wholly qualitative approaches with children are rarer and limited to a handful of published studies and descriptive sessions and vignettes by psychotherapists in the field (Brooks, 2006; Chardonnens, 2009; Karol, 2007; Vidrine, Owen-Smith & Faulkner, 2002), although there is more 'grey' literature in the form of theses and dissertations (see e.g. Esbjourn, 2006; Frame, 2006; Hayden, 2005). It is hoped, therefore, that this book will provide a contribution to this knowledge base and make a useful addition to the research concerned with the qualitative, subjective processes involved in, and young peoples' own understandings of, TH, rather than with measuring outcomes. Nevertheless, it is accepted that an outcome focus can provide useful additional information and is necessary for credibility and funding in the field.

Animals and society

Humans and non-human animals have occupied and shared their lives and spaces throughout the history of humankind. Whilst there are a number of competing theories behind the history of human and non-human animals, ranging from biological, Darwinian based theories to more recent cultural and sociological understandings of animals in the lives of humans, it is claimed that

> it is now being fully recognised at the analytic level that animals are crucial to the functioning of any society, in that they provide for

8 *Equine-Assisted Therapy and Learning with At-Risk Young People*

humans food, labour, raw materials, modes of transportation, companionship, scientific knowledge through observation and experimentation, and forms of leisure and entertainment amongst other things.

(Wilkie & Inglis, 2007: 3)

In addition to the roles that they play in our lives outlined above, animals, and especially horses, have been included in stories, myths and legends throughout the ages and across cultures (Chamberlin, 2007; Davies & Jones, 1997; Howey, 2002; Jackson, 2006; Runnquist, 1957; Walker, 2008). These complex arrays of relationships are fraught with as many contradictions as there are in society. On the one hand, animals are exploited, dominated, eaten, worn and experimented on in medical science, and, on the other, the same species are kept as pets, used for therapy, seen as a way for humans to have contact with nature and wilderness, and used as a way of defining who we are. In order to attempt to deal with some of these complexities, industrial and post-industrial modern societies have evolved processes where the messy parts of the lives of many animals have become invisible, especially in relation to food production, with the lives and deaths of food animals, such as those in the factory farm and slaughter house, largely being hidden away from people's lives. Most urban dwellers are distanced from wild or farm animals, but conversely 'pet' animals have become more visible. There are confusing contradictions in these relationships in that some animals, the pets, are seen as fluffy, cute and become anthropomorphised, in contrast with how other (and the same) animals are raised and farmed for food (both for humans and their pets), fur and experimentation, as well as entertainment in zoos, sports and circuses (Wilkie & Inglis, 2007).

Animals as 'good to think with'

Some critics have argued that sociologists have largely neglected studying the role of animals in society due to the historical anthropocentric belief that only human animals matter (Serpell, 1999; Wilkie & Inglis, 2007). Animals in the lives of children have been almost entirely ignored by sociologists (Melson, 2001; Serpell, 1999), although the place of animals in society and culture has had greater prominence in anthropology and an emerging animal geography (Arluke & Saunders, 1996; Franklin, 1999; Philo & Wilbert, 2000). Drawing on the earlier work of Levi-Strauss (1968), these disciplines suggest that 'animals offer a window into human thinking and needs' (Arluke & Saunders, 1996: 3). Because animals are both like us and not like us, they are uniquely positioned to provide us with the opportunity to learn more about ourselves because 'studying animals and human interactions with them enables us to learn about ourselves as social creatures' (Arluke & Saunders, 1996: 4). Philo and Wilbert suggest that animals are responsible for having helped to shape our cultures because

> Humans are always and have always been, enmeshed in social relations with animals to the extent that the latter, the animals, are undoubtedly constitutive of humans societies in all sorts of ways.
>
> (2000: 2)

These ideas are linked to some of the literature on how animals and nature are used in symbolism, and the ways in which these symbols can represent some of our emotional states, feelings and other attributes. Many words and phrases which employ horses, other animals and nature can be used to illustrate this. For example, racehorses are described 'as fast as the wind' – a reference to the Greek myth of Achilles and his marvellous horses, which were said to have been born from the West Wind (Howey, 2002). Other terms are 'wild horses wouldn't keep me away', 'sly as a fox', 'courage of a lion', 'free as a bird', 'slippery snake' and many more. In Jungian psychology, animals are often seen as representing certain aspects of ourselves in dreams, as being 'innate' archetypes held within the 'collective unconscious', although there are different perspectives on Jung's interpretations of these concepts (Jones *et al.*, 2008; Jung, 1978; Knox, 2003).

Differences between humans and non-human animals

Historically, how a society treats non-human animals has been used by philosophers, storytellers and artists in an attempt to explore moral issues and greater philosophical questions of who we are and why we are here (Kemp, 2007). By determining that human animals are different from non-human animals because humans possess a 'rational' self-conscious mind and are 'civilised', as opposed to 'uncivilised' animals that behave on instinct and primal needs alone, animals have been used to create an understanding of what it is to be human (Philo & Wilbert, 2000: 14).

Big questions of whether animals possess feelings, souls and consciousness, or are machines, as Descartes declared, have accompanied our treatment of animals over time (Kemp, 2007). In feudal England, humans and animals lived in close proximity to each other and 'during winter it was common throughout Britain for people and beast to share the same roof' (Franklin, 1999: 11). It has been suggested that because of this close proximity to and relationships with animals, it was necessary to create a division between ourselves and our non-human co-habitants in order to have some boundaries and separation, otherwise we were all in danger of being the same, and humans would have no superiority (Franklin, 1999; Philo & Wilbert, 2000). Christianity provided the legitimisation of human–animal difference as religion maintained that God positioned 'man' as dominant over nature and animals: 'Man being made in the likeness of God' (Wilkie & Inglis, 2007: 8). This view was substantiated by Saint Augustine (354–430), who claimed that animals do not possess souls and were therefore vastly inferior to humans, so animal suffering and cruelty were acceptable. Because they were not seen as possessing consciousness or having a soul, it became easier to believe that animals do not feel pain like humans and brutal practices towards them, such as bear-baiting and cock-fighting, became an acceptable and necessary means of creating a distinction and distance between animals and humans (Franklin, 1999; Tester, 1991). It is obviously easier to morally justify eating, exploiting and carrying out scientific experiments on another living being if it is considered different and inferior, and does not possess feelings (Bekoff, 2007). This dominant discourse began to be challenged in the mid-nineteenth century with the Enlightenment and Romantic movements, Jeremy Bentham asserting that 'if all pain is evil then pain and suffering caused to an animal by a human must be evil despite the absence of rationality in the animal's mind' (cited in Wilkie & Inglis, 2007: 8). A romantic view of nature emerged, and the beginnings of the animal rights movement arose alongside the rise of industrialisation and the movement away from a life with animals in the country to modern, urban living, distanced from nature and animals. The growth of capitalist society brought further changes to our relationships with animals: more leisure time together with mass consumerism gave rise to animals and nature becoming profitable marketable commodities (Franklin, 1999; Louv, 2008; Roszak, Gomes & Kanner, 1995). An increased interest in nature conservation, zoos, national parks and wildlife theme parks alongside hunting and fishing as leisure and sports activities add further complex contradictions to how we relate to and live with animals (Franklin, 1999; Haraway, 1991, 2008). Postmodern thinkers brought a deeper understanding that a history of exploitation of animals follows

Background to Equine-Assisted Therapy and Learning   11

a similar trajectory to other marginalised and discriminated groups in society. The history of slavery and women's rights together with the acknowledgement that ethnic minorities and people with disabilities have suffered discrimination and been judged as inferior to the 'cultured man', the white male, are compared to the history of the treatment of animals by some authors (Birke, 1994; Dunayer, 2004; Haraway, 1991, 1992). Language has been employed accordingly to reinforce this notion, with words such as 'bitch', 'cow', 'jackass', 'beastly', 'brutish', 'monkey' and so on becoming terms of repression, and, 'Like sexist language, speciesist language fosters exploitation and abuse' (Dunayer, 2004: 11).

New questions about our relationships with animals are raised with recent scientific and medical developments on the use of animals in research. As humankind has, on the one hand, begun to blur the boundaries between animal, machine and human, and animals used in science and technology have become distanced further away from our daily lives, they are, simultaneously, potentially becoming part of our bodies with the 'advances' in medical science now making the very real possibility of animals being farmed for their organs for use in human transplants. The fusion of human and machine has been raised as having important implications by some authors, with the division between human, animal and robot becoming blurred (Haraway, 1991, 1992; Louv, 2008; Mazis, 2008). The creation of chimeras and microscopic 'nanorobots' capable of living deep inside our cells is now a reality and 'the boundary between human and animal is thoroughly breached' (Haraway, 1991: 151). A further paradox is that whilst these particular animals become more 'othered' in order to become part of us, there is a burgeoning growth in the use of animals in an emotional and therapeutic context within the field of AAT (Chandler, 2005; Dossey, 1997; Fine, 2000; Friedmann *et al.*, 1983; Levinson, 1969, 1980; Podberscek, Paul & Serpell, 2000). This corresponds with the recent (re-)emergence of a move to get 'back to nature', and ecological awareness and concern. The fields of nature and wilderness therapy and ecopsychology, with their roots in the sometimes controversial work of E.O. Wilson's biophilia theory (1984), believe that humans are intrinsically, biologically, related and part of nature, and consequently need contact with the natural environment in order to flourish and grow (Devall & Sessions, 1985; Kellert & Wilson, 1993; Louv, 2008; Roszak, Gomes & Kanner, 1995; Shepard, 1982; Wilson, 1984). The AAT field has a number of links to the biophilia hypothesis and this is explored in more depth later in this chapter.

Children and animals – a sociological perspective

Whilst there is limited literature on EAT/L, there is even less on the relationship between children and animals within the fields of sociology, psychology and child development. Serpell observes that 'the sad truth is that psychologists and social scientists have shown a baffling lack of scholarly interest in the child-animal relationship' (Serpell, 1999: 92).

The research and literature agenda around animals has primarily been concerned with animal rights, ethics and animals in research (Arluke, 1994; Bekoff, 2007; Regan, 1983; Singer, 1990), how we think about animals and animals in culture (Arluke & Sanders, 1996; Franklin, 1999; Haraway, 1992, 2004), and our connection with animals (Bekoff, 2007; Irvine, 2004; Serpell, 1999, 2000a; Podberscek, Paul & Serpell, 2000). As previously stated, children are almost entirely absent from this literature with a few notable exceptions, such as Melson (2001) and Myers (2007). The area of AAT has looked a little more in depth at the therapeutic and health benefits of contact between animals and children, with the often cited study of how the blood pressure of children was lowered when they stroked dogs (Friedmann *et al.*, 1983), but it is still largely adults who dominate the AAT literature (Fine, 2000; Irvine, 2004; Serpell, 1999). This is perhaps linked to the long-standing Western tradition of denying the voice of children and giving them only limited rights (Alderson, 2001, 2004; Christensen & Prout, 2002, 2005; James & Prout, 1990). Myers takes this theme further, suggesting that Western culture has a deep-rooted tradition of viewing children as essentially unformed and 'animal-like' (Myers, 2007). In her study on children's lives with animals, Melson (2001) shows surprise at how the field of child psychology and development has ignored the roles of animals, and especially pets, in children's lives, considering the prominence of attachment and transitional object theories (Bowlby, 1984, 1988; Winnicott, 1953, 1965). She proposes that animals provide much opportunity for children to experience healthy attachments and to explore transitional experiences through animals and play, mentioning Freudian and Jungian theories of animal symbolism and suggesting that animals are useful because 'children readily access animals as material in the development of a sense of self' (Melson, 2001: 20). Melson suggests that in addition to their role in child development, animals offer many aspects that have been largely ignored by scholars writing about children, arguing that the study of children has been 'humanocentric'. She points out that pet-keeping is 'apparently universal across human groups and so old it co-evolved with modern humans', proposing that this is why animals are so prominent in children's stories, myths, rhymes and fairy tales (Melson, 2001: 14). Examples given of this include 'Goldilocks and the Three Bears', 'The Ugly Duckling', 'Little Red Riding Hood' and so on. It is argued that the use of animals continues in contemporary children's literature, in films, cartoons, toys, and in marketing, with Bambi, 101 Dalmatians, Mickey Mouse, Donald Duck and the Andrex puppy being just a few examples (Melson, 2001; Serpell, 1999).

Melson goes on to express the view that animals are useful in teaching responsibility to children, together with a valuable healing role through providing companionship, emotional support and opportunity to experience love and affection (2001, 2003). She goes on to hint that for some children, animals are more important to them than the people in their lives and that 'for many children ... pets are more likely to be part of growing up than are siblings or fathers' (Melson, 2001: 34). In addition is their role as 'social lubricant' (Corson & Corson, 1980; Fine, 2000; Wells, 2009) and their ability to enhance the therapist–child relationship, as initially suggested by the child psychologist and pioneer of AAT, Levinson (1969), and later Fine (2000). Other authors exploring the role of animals in the family from a social work perspective support Melson's conclusion that animals are important to children who are suffering from trauma and distress, and that animals have the capacity to offer healthy attachment experiences (Tedeschi, Fitchett & Molidor, 2005; Turner, 2005; Walsh, 2009) and opportunities for the acquisition of empathy (Poresky, 1990).

AAT

A small pet is often an excellent companion for the sick, for long chronic cases especially.
(Florence Nightingale, 1860, cited in Dossey, 1997: 8)

The pleasure that people appear to receive from living with animals would seem hard to dispute, with 26% of the population of the UK owning dogs and 31% cats, according to a random sample of 2980 households (Murray *et al.*, 2010). Their value as a therapeutic intervention is also acknowledged, with guide dogs for the blind being the most obvious example, although pet-assisted therapy (www.petsastherapy.org), an intervention mainly concerned with taking pets to visit people in hospitals and hospices, is now expanding in the UK. In addition there are therapy programmes where people swim with dolphins, although these are not without criticism (Chandler, 2005), and 'donkey-facilitated therapy', where donkeys are taken into residential homes for the elderly and employed in therapy with children (www.elisabethsvendsentrust.org.uk). In the USA, where AAT is well established and has a longer history, there is a larger research base (Barker & Dawson, 1998; Beck & Katcher, 1996; Bernstein, Friedman & Malaspina, 2000; Chandler, 2005; Fine, 2000). It is claimed that animals can offer benefits to people with a range of health issues, from physical to mental health problems (Anderson, Hart & Hart, 1984; Chandler, 2005; Dossey, 1997; Fine, 2000; Levinson, 1969; Podsberscek, Paul & Serpell, 2000; Serpell, 1999, 2000a; Siegel, 1993). Interest in the field is rapidly expanding with the establishment of the Delta Society in the USA and the Society for Companion Animal Studies (SCAS) in the UK. These and other organisations, such as veterinary pharmaceutical companies, have helped to fund the undertaking of research into the subject.

Background

Early records of the employment of AAT have been traced to the York Retreat, an asylum founded in 1792 by the Quaker movement. 'Bethel', a home for epileptics in Germany, followed in 1867, both centres keeping animals for the patients to interact with and care for (Melson, 2001; Netting, Wilson & New, 1987). Whilst early reports of the success of AAT were anecdotal and mainly consisted of records by psychotherapists (Chandler, 2005; Levinson, 1969; Serpell, 2000a), research into AAT over the past ten years or so has largely aimed towards adopting a positivist, quantitative approach utilising empirical, experimental methods to test the hypothesis that AAT is effective in promoting health in many different respects (Hooker, Freeman & Stewart, 2002). A key study to have caught the public's attention claims to show that stroking a dog reduces stress and lowers blood pressure (Friedmann *et al.*, 1983). The previous influential study by these authors showed that pet owners had improved survival rates among patients a year after suffering from a heart attack (Friedmann *et al.*, 1980). Many other studies and publications followed, supporting similar and additional health claims (Anderson, Reid & Jennings, 1992; Beck & Katcher, 1996; Bernstein, Friedman & Malaspina, 2000; Mallon, 1994; Marr *et al.*, 2000; Serpell, 1991; Siegel, 1993).

Animals and pets for health and therapy

It is suggested that pets and other animals can offer further psychosocial benefits in addition to the physical health claims of pet ownership.

Siegel (1993) claims that pet owners are happier than non-pet owners and subsequently experience fewer physical problems, measuring this via a study that demonstrated that pet owners experienced fewer visits to their GP compared with non-pet owners over a one-year period. Other studies make similar physical health claims (Headey, 1999; Rainer *et al.*, 1999). One reason proposed for this is that because dogs, for example, need exercise, keeping animals can encourage physical activity, the health benefits of exercise on both physical and mental health being widely recognised (Halliwell, 2005; Munoz, 2009; Peacock, Hine & Pretty, 2007). If animals can provide motivation for the previously unmotivated to exercise, then mental health and physical health may therefore be improved. In addition to claims about the benefits of animals for general health, reported psychosocial benefits of AAT range from animals providing relationships and support (Beck & Katcher, 1996; Risley-Curtiss, 2010; Walsh, 2009; Wells, 2009; Wilson & Turner, 1998) to encouraging responsibility, empathy and moral development (Daly & Morton, 2006; Myers, 2007; Poresky, 1990), and growth of self-esteem and control of behaviour (Fine, 2000; Kogan *et al.*, 1999). Other psychological theories include seeing the animal as a transitional object (Trienbenbacher, 1998) and acting as 'confidant' and 'holding environment' (Brooks, 2006; Melson, 2001). Many authors make links to pets and attachment theory with both children and adults (Beck & Madresh, 2008; Crawford, Worsham & Swinehart, 2006; Parish-Plass, 2008; Sable, 1995; Tedeschi, Fitchett & Molidor; 2005; Walsh, 2009). In addition, it is claimed that pets provide companionship and so decrease loneliness, provide security and comfort, fulfil the need for tactile stimulation and serve as non-judgemental support (Chandler, 2005; Dossey, 1997; Podsberscek, Paul & Serpell, 2000; Serpell, 1999, 2000a). The psychologists Levinson (1969) and Corson and Corson (1980) wrote about how they found animals to be powerful facilitators in psychotherapy with both children and adults. Another suggestion is that animals can be especially helpful for those who find 'normal' social interaction difficult as they can be 'social lubricants', acting as a catalyst for people to make contact with others in an informal, non-threatening way. Corson and Corson in their study suggest that pets offer isolated and withdrawn people

> a form of non-threatening, non-judgemental, reassuring nonverbal communication and tactile comfort and thus helped to break the vicious cycle of loneliness, helplessness and social withdrawal.
>
> (1980: 107)

A further theory put forward is that because of the responsibility required to look after an animal, self-esteem is increased, and it is claimed that children who have contact with animals have increased social competency (Edney, 1995), increased social skills (Katcher & Wilkins, 2000) and empathy (Daley & Morton, 2006; Paul & Serpell, 2000). Similar claims are made about AAT provided in prisons, and about a reduction in violence and antisocial behaviour and recidivism rates when animals are included in prison programmes (Strimple, 2003).

It has been argued that these benefits of AAT are due to a deep connection between animals and humans related to the biophilia concept (Katcher & Wilkins, 2000; Levinson,1972; and Serpell, 2000b), and that animals can 'help to connect or reunite people with something fundamental within themselves – a sort of unconscious "animal within"'(Serpell, 2000b: 110). Other authors claim that we have an especially deep connection with horses, over and above other non-human animals, due to our long history and modern civilisation being based on our relationship with them over millennia (Barclay, 1980; Chamberlin, 2007; Walker, 2008). In addition, it is claimed that horses have long been considered to have supernatural and healing powers throughout many cultures, providing further evidence of their effectiveness in the fields of EAT/L (Baker, 2004; Kohanov, 2001; McCormick & McCormick, 1997, 2004).

History of riding therapy and EAT/L in the present day

It is widely considered that horses have played a fundamental part in the history of humankind. Cave paintings dated around 13,000 BC show horses being hunted as food and for their hides, whilst it is believed that they became domesticated and were first ridden around 1500 BC in Central Asia, although there is debate about the exact date (Hallberg, 2008 Walker, 2008). This led to the crucial roles that horses have played in war and transport, and they subsequently became deeply implicated in our modern societies (Barclay, 1980; Budiansky, 1998; Chamberlin, 2007; Hyland, 1996). The long-forged relationship between human and horse has resulted in the horse being ingrained in the human psyche (Game, 2001; Frewin & Gardiner, 2005; McCormick & McCormick, 1997) in a similar way to the innate link between humans and nature claimed by biophilia theory (Wilson, 1984). In turn, it is argued that horses can be employed therapeutically to awaken lost aspects of our consciousness (Held, 2006; Kohanov, 2001; McCormick & McCormick, 1997, 2004). These authors suggest that long before the horse's unique characteristics were harnessed by humankind and enabled our evolution as a species, they were believed to possess magical, healing powers. Cave paintings provide illustration that horses were revered as far back as 20,000 years ago, palaeontologists suggesting that the paintings in Pech-Merle in France indicate that early man thought the horse had magical powers in warding off evil spirits and protecting against disease and danger (Chamberlin, 2007; Mayberry, 1978). Riding as a therapy can be traced back to the fifth century BC, with it being suggested that there is evidence demonstrating that wounded Greek soldiers were encouraged to ride horses to increase their morale (All, Loving & Crane, 1999; Barclay, 1980; Cawley, Cawley & Retter, 1994). Greek mythology has especially strong symbolism of the horse as a healer. Examples are Pegasus, who was thought to bring luck, inspiration and knowledge, and Chiron, the chief centaur, who was the god of healing, the word 'chiropractor' having originated from Chiron. It is from the Greek philosopher Hippocrates that the medical tradition emerged – *hippo* being the Greek word for horse, and so underpinning the link between horses and healing within Greek culture (All, Loving & Crane, 1999; Broersma, 2007). The Greeks continued to play an important part in horse history: classical riding and dressage have roots in the work of Xenophon, the important trainer and philosopher who believed in cultivating the partnership between man and horse (Chamberlin, 2007; Loch, 1999). This reverence towards the horse has found its way across cultures and continents. In Iberia the horse was believed to have come out of the sea from the lost city of Atlantis as the 'Son of the Wind', an ancestor of the unicorn who, again, like Pegasus, was thought to bring luck (Broersma, 2007; McCormick & McCormick, 1997). North American Indians, who have long been admired for their natural riding ability, seemingly being at one with their horse, recognise them as healers and spiritual guides, as do the Mongolian Shamans (Baker, 2004; Isaacson, 2009). To the ancient Celts, horses were sacred creatures and were associated with the life cycle, power, fertility, and physical and metaphysical travel, together with good fortune. They believed that the horse goddess was responsible for transporting the human soul into and out if its earthly existence at the appropriate season of life (Chamberlin, 2007; Davies & Jones, 1997; McCormick & McCormick, 2004).

Horses and therapy in the present day

Fascination with the horse and its embedment in society as symbolic and cultural icon would appear to be just as strong in the present day, with horses represented in fairy tales, films, books, art and poetry along-side their role in sport and recreation (Barclay, 1980; Chamberlin, 2007; Jones, 1983; Lawrence, 1984; Scanlan, 2001; Walker, 2008). Horses have long been represented in symbolism, in magic and mystery, legend and myth. They are often considered to be lucky in superstition and historically were considered to be healers in many cultures (Howey, 2002; Jackson, 2006; McCormick & McCormick, 2004). The horseshoe is still a lucky icon and appears on cards and cakes to celebrate christenings and birthdays, and it is thrown as confetti at weddings. White horses (considered to be good luck) are often employed at weddings, either ridden by the bride or drawing a carriage, and black horses pull funeral hearses (Jackson, 2006). It has been suggested that the horse offers the unique combination of being both extremely powerful and potentially dangerous, but also being gentle and a 'symbol of human spirit and freedom' (Frewin & Gardiner, 2005: 4). As our lives have become safer and more sedentary, it has been argued that horses fulfil an innate need to take risks, and perhaps this is one reason why it is within sport and recreation that they have found their niche in modern society (Rosenthal, 1975). Undoubtedly, horse-related sports have their roots in war and from Greek chariot racing. The hunting of wild boar and deer by lords, kings and the aristocracy followed, which led to modern-day fox-hunting (Barclay, 1980; Chamberlin, 2007). Even more often, however, the horse is a symbol of power, which of course can represent freedom, as the following extract from a passage from the Book of Job (xxxix. 19–25, A.V.) illustrates:

> Hast thou given the horse strength? Hast thou clothed his neck with thunder? Canst thou make him afraid as a grasshopper? The glory of his nostrils is terrible. He paweth in the valley, and rejoiceth in strength: he goeth on to meet the armed men. He mocketh at fear, and is not affrighted ... He swalloweth the ground with fierceness and rage...

A theme of how horses represent freedom and wildness on some level is something that the anthropologist Elizabeth Lawrence found in her study on rodeo, which explores how horses are experienced and seen as both wild and yet also tame and, additionally, how the horse 'serves as a symbolic bridge between nature and culture' (Lawrence, 1984: 132). She suggests that the horse's power and inherent wildness represents humans' domination over nature in some way, claiming: 'I suggest the horse as an archetypal symbol of man's conquering force' (Lawrence, 1984: 134).

It is from sport that the therapeutic value of horses has emerged, although this was until recently largely based on the physical benefits of riding due to the inspiration sparked by Lis Hartel, the world-renowned dressage rider of Denmark. She won a silver medal at the 1952 Helsinki Olympic Games having taken up riding again after being severely paralysed by poliomyelitis (All, Loving & Crane, 1999; Davies, 1988; Kaiser *et al.*, 2004). Inspired by both the physical and the psychological benefits that Hartel derived from riding, physiotherapists at the orthopaedic hospital in Copenhagen began to prescribe riding for patients. News of its success as a treatment spread to Britain, where the RDA was formed in 1970, and which now has national member organisations worldwide (Davies, 1988). The value of riding as a therapeutic tool for the physically disabled has been well documented and there are numerous studies that have been conducted to attempt to empirically prove its physical benefits, due to the fact that

> riding provides the person with a disability a normal sensorimotor experience that contributes to the maintenance, development, rehabilitation, and enhancement of physical skills. The sensorimotor experience involves vestibular input which stimulates the rider's balance mechanism as constant adjustments are needed with the horse's movement at different gaits.
>
> (All, Loving & Crane, 1999: 52)

Pioneering research in the 1980s by Bertoti (1988) on the benefits of hippotherapy for children with cerebral palsy showed significant improvement in posture following a ten-week programme. Improvement in coordination was reported by Brock (1988) in his study of 15 adults with various physical disabilities, whilst MacKinnon *et al.* (1995) found that there were significant improvements in posture, trunk control, attention span, pelvic mobility and hand control in their study of 19 children aged between four and twelve who had been diagnosed with spastic type cerebral palsy.

Studies in EAT/L

Although it is steadily increasing, there is still relatively little research on EAT/L, with many of the studies to date which concentrate on aspects of the psychosocial benefits of EAT/L being qualitative, case-study and anecdotal accounts by practitioners in the field. Kemp *et al.* (2013) warn of a need to be cautious due to the limited number of published empirical studies, with Bachi going further and stating that 'existing knowledge about this field is insufficient, and most of the research suffers from methodological problems that compromise its rigor' (2012: 364). This includes a lack of control groups, small sample sizes and the limited duration of the studies conducted (Bachi, 2012; Kemp *et al.*, 2013; Selby & Smith-Osborne, 2013; Smith-Osborne & Selby, 2010).

A review of notable studies that explore how the horse can provide beneficial therapeutic and/or learning experiences finds that this can be in a variety of ways. For example, the horse is described as being non-judgemental and motivational (Bowers & MacDonald, 2001; Yorke *et al.*, 2008), useful as metaphor (Karol, 2007; Klontz *et al.*, 2007) for building self esteem, confidence and mastery (Bizub *et al.*, 2003; Burgon, 2003; Trotter *et al.*, 2008; Vidrine, Owen-Smith & Faulkner, 2002), adapting behaviour (Kaiser *et al.*, 2004; Shultz,Remick-Barlow & Robbins, 2007), effective for building trust and attachment with both the horse and therapist (Brooks, 2006; Chardonnens, 2009; Halberg, 2008; Yorke *et al.*, 2008), and useful in grief counselling (Symington, 2012). In the USA, equine therapy-based programmes have been utilised for some years in the rehabilitation of offenders (Bachi, 2013b). An early study in this area was interested in exploring the participants' own meanings of their interactions with horses and looked at the benefits of horses to inmates at a prison in New Mexico, utilising both qualitative and quantitative methods. The Wild Mustang Program (WMP), as it was known, claimed a reduction of disciplinary reports and recidivism. This was 'particularly for those who simultaneously participated in substance abuse counselling and for violent offenders' (Cushing & Williams, 1995: 95). The rehabilitation scheme comprised 69 inmates evenly divided between those convicted of violent and property crimes with an average age of 35. The objective of the programme was for the participants to each tame a wild mustang within the time limit of a month. Questionnaires were then given to both the correction officers and the inmates to gauge what the perceived benefits were. Recidivism rates were also analysed and compared with the general recidivism rates of New Mexico correctional facilities. They were shown to be significantly lower for the WMP: 14.82% in comparison with the general rate of 38.12%. Although the programme was administered along different lines than the way most general therapeutic riding and EAT/L programmes have developed, in that the horses were wild and were 'broken in' – in the Western sense, often a violent encounter (for both the participants and the horses) – it nevertheless served as a therapeutic experience for many of the inmates as demonstrated by

Background to Equine-Assisted Therapy and Learning 21

their testimonials, which reported improvements in self-confidence and development of empathy. The authors report that there was a 29% reduction in disciplinary reports during and after the programme participation and this was especially high for major reports, with a 55% reduction.

In their study with young people, Vidrine, Owen-Smith and Faulkner (2002) also employed a descriptive, qualitative approach to look at an intervention employing vaulting as equine-facilitated group psychotherapy for children from a range of backgrounds, including foster care, residential treatment settings and school problems. These sessions involved a number of children aged between seven and ten years participating in vaulting sessions (learning a set sequence of controlled, gymnastic moves whilst on the back of a horse). The sessions also involved horse care and understanding horse behaviour, as well as learning to work in a team. Participants attended sessions for 90 minutes a week for 8–10 weeks. The authors provide a descriptive account of the vaulting sessions, employing a participant-observation model and then interpreting the process by using their own analysis in addition to statements from the children, their carers and other professionals involved. Two of the founders of the centre, HorseTime in the USA, were a clinical psychologist and a psychiatric mental health nurse, and the study is framed within a psychotherapeutic approach drawing on experiential therapy with its roots in the psychodynamic tradition of Jung.

One of the themes that the authors drew from the study is that the horses facilitated healthy physical tactile experience, stating:

> As clients became more comfortable in the farm environment, they were noted to be appropriately physically affectionate with both the horse and the staff. Even the boys who were seldom seen expressing physical affections with humans were observed hugging and kissing their horse.
>
> (Vidrine, Owen-Smith & Faulkner, 2002: 600)

They also described how trust was built, both between the participants themselves and with their horses. 'By the end of the group, however, nearly every child had widened their circle of trust' (Vidrine, Owen-Smith & Faulkner, 2002: 599). The authors conclude that therapeutic vaulting

> offers a unique opportunity for experiential group psychotherapy experience. Participants are able to address developmental, personal

22 *Equine-Assisted Therapy and Learning with At-Risk Young People*

and social needs in the context of a somatically engaging, challenging and enjoyable activity.

(Vidrine, Owen-Smith & Faulkner, 2002: 602)

However, whilst Vidrine, Owen-Smith and Faulkner claim that the
above benefits were apparent from their study, they go on to suggest that
more research 'across a variety of treatment settings and with a variety
of clinical populations' is needed in order to establish the effectiveness
of EFP (2002: 602). The study had some flaws in that the authors did
not specify how many young people took part in the study or for how
long the study was conducted, and they did not discuss any other factors, such as staff, choosing instead to focus on providing a descriptive
account of a typical group.

Utilising a similar approach to the research study on which this
book is based, but with adults as opposed to young people, a study by
Bizub, Joy and Davidson (2003) followed the experiences of five adult
participants in a therapeutic horse-riding programme over ten weeks.
Employing a case-study participant observational model of enquiry, the
researchers from Yale University carried out this study in conjunction
with an established therapeutic riding and EAT centre in the USA. The
participants, who volunteered for the programme, were adults with
long-standing mental health problems, primarily within the schizophre-
nia spectrum. In addition to riding, the emphasis of the programme
was on bonding exercises with the horses and a post-session processing
group. The researchers, who were psychology graduates and practitioners, co-facilitated the sessions each week, so demonstrating the
participant observational approach of the study. Another data-collection
method employed consisted of conducting semistructured interviews
with the participants. The epistemological framework of the study can
be demonstrated by the authors' concerns about ensuring that the participants' voices were heard and of the methods employed being an
empowering experience for the individuals concerned. In the analysis
of the programme they write:

When one enters the system of psychiatry, there is the transformation
of one's rich 'personal story' into a 'case history'. One way to counteract this disempowering occurrence is for people with disabilities to
tell their own stories and be listened to. Thus qualitative analysis was
used to describe how, and in what way, if at all, this programme was
transformative for participants. By inviting individuals to describe

their personal experiences so as to grasp their meaning, this method is inherently empowering.

(Bizub, Joy & Davidson, 2003: 380)

A discussion of the results was presented in the participants' own words, with the themes that emerged being explored. These ranged from the riders describing a growth in confidence from learning a new skill to overcoming the fear associated with riding a horse for the first time. The participants began to gain insights into themselves through this, and to apply them to other areas of their lives, 'because I was able to overcome a fear, that helped me' and 'I know that they [the horses] get scared and they run and I found out it's just like me' (Bizub, Joy & Davidson, 2003: 381). Another key theme to emerge was that of the horse offering 'unconditional love' and of the participants building up strong relationships with the horses that then helped to initiate other social connections. The authors concluded that this was due to it being a 'normalizing experience' (Bizub, Joy & Davidson, 2003: 381). They liken therapeutic horse-riding to exercise therapy and outdoor adventure activities, but with the additional dimension of the horse offering further opportunities for rehabilitation, writing that

Getting in touch with a horse, both literally and figuratively, can enhance one's appreciation for both the horse and one's self...we suggest that therapeutic horseback riding is a promising way for such individuals . . . to more fully participate in their own lives.

(Bizub, Joy & Davidson, 2003: 383)

Another study on the benefits of EAT/L measured psychological well-being of adults before and after participating in equine-assisted experiential therapy, finding positive outcomes (Klontz et al., 2007). Other studies claim that therapeutic riding can significantly reduce anger issues with children (Kaiser et al., 2004, 2006), and another looked at EAT/L therapists' perspectives on the interventions in order to gain an overview of the perceived benefits (Young & Bracher, 2005). In their study with 14 at-risk adolescents, Bachi, Terkel and Teichman (2012) examined the outcomes of EFP on self-image, self-control, trust and general life satisfaction. The study was conducted over seven months on a weekly basis and included a control group. Whilst the authors conclude that their hypothesis that the young people participating in the EFP programme would show a greater improvement in the four areas of study than the control group was not statistically significant, they show that the data did indicate a trend in this direction and the qualitative descriptions support this (Bachi, Terkel & Teichman, 2012).

Ewing *et al.* (2007) and Schultz, Remick-Barlow and Robbins (2007) also employed both quantitative and qualitative methods to evaluate EAP and equine-facilitated learning (EFL) programmes for children and young people with behavioural and mental health problems. Both studies employed pre- and post-testing, with Schultz, Remick-Barlow and Robbins (2007) finding improvements in GAF (Children's Global Assessment of Functioning Scale) scores for all the children participating in EAP. However, in the EFL study by Ewing *et al.* (2007), no statistically significant improvement was found in the measurement of self-esteem using the Harter Self-Perception Profile measuring technique, although, conversely, qualitative, observational data collected did report 'positive changes in conduct and social acceptance' (Ewing *et al.*, 2007: 68). The authors report having difficulties in administering the psychological testing with this group of children, stating that 'many children had difficulty understanding the questions asked of them during testing' and 'several students would begin to display disinterest or agitation to end the testing period' (Ewing *et al.*, 2007: 70). This was something also encountered in this research study with a number of the young people when asking questions, as will be seen in the following chapters.

Studies of EAT/L are still in their infancy and are fraught with methodological, ethical and epistemological difficulties and differences, not least when looking at vulnerable and excluded children and young people within the care system, together with the multitude of challenges and variables that working with both horses and young people brings. Despite this, the handful of studies that have been conducted so far, in the majority, appear to show that 'Equine-assisted psychotherapy (EAP) is fast gaining recognition internationally as an effective treatment strategy for a number of different client groups' (Frewin & Gardiner, 2005: 2).

It will be seen that many of the themes and therapeutic benefits claimed by the studies introduced in this chapter arise in this book: themes of a growth in confidence and self-esteem gained from learning a new skill and overcoming fear (Bizub, Joy & Davidson, 2003; Cushing & Williams, 1995), and themes of attachment and healthy tactile stimulation (Bachi, 2013a; Vidrine, Owen-Smith & Faulkner, 2002), together with the opportunity to change behaviours and build positive therapeutic relationships (Bass, Duchowny & Llabre, 2009; Ewing *et al.*, 2007; Kaiser *et al.*, 2004; Kemp *et al.*, 2013; Rothe *et al.*, 2005).

Why horses?

How far animals can be a useful therapeutic intervention for children and adults with additional emotional and social needs has been demonstrated to be the subject of ongoing discussion and research. In addition, the relationship between people and horses is the subject of different theories and perspectives. As briefly explored earlier in this chapter, these range from how horses can be powerful facilitators for healing within a spiritual dimension (Baker, 2004; Broersma, 2007; Kohanov, 2001, 2005; McCormick & McCormick, 1997, 2004) to a historical exploration (Barclay, 1980; Chamberlin, 2007; Hyland, 1996). The importance of their role in our sociological and cultural space is explored by Birke (2007, 2008), Latimer and Birke (2009) and Lawrence (1984), with Latimer and Birke arguing that horses are 'deeply implicated in the production and reproduction of culture and society' (2009: 3). Brandt (2004) looks at the process of human–horse communication and how shared meaning is created, whilst traditional biological theories together with natural horsemanship methods of horse behaviour are discussed by, amongst others, Budiansky (1998), McGreevy (2004), Rashid (2004), Rees (1984) and Roberts (2000).

Interestingly, horses occupy a space 'in between' other animals. They are neither quite a pet nor classed as farm animals, and this is reflected in law in that horses, although living in agricultural spaces, are not included in legislation covering the welfare of agricultural animals (DEFRA, 2006). Horses are in our lives in an array of often contradictory ways, as work animals, sporting partners, leisure and companion animals, symbols of wilderness and freedom, and therapeutic 'tools'. Yet horsemeat is eaten in many countries. Reasons put forward for why the horse can offer a unique dimension to the therapeutic arena are varied, but one theory suggested by practitioners in the EAT/L field is that the horse is especially effective in psychotherapeutic work because it 'is an animal of great power and grace: yet it is also one of inherent vulnerability' (Karol, 2007: 78). This vulnerability is due to the horse being a very tasty herbivore that has had to develop highly honed communication skills within its herd to survive over millennia (Budiansky, 1998; McGreevy, 2004; Rashid, 2004; Rees, 1984). Whilst there are competing perspectives on the traditional understanding of herd structure and dominance theories (Berger, 1986; Houpt, Law & Martinisi, 1978; McGreevy, 2004; Rees, 2009; Roberts, 2000), a growing number of horse trainers and ethologists in the field of natural horsemanship suggest that horses are inherently cooperative animals

26 *Equine-Assisted Therapy and Learning with At-Risk Young People*

and fine-tuned to the body language and emotions of both people and horses:

> the thing to remember is that horses are, by nature, very co-operative and social animals. They have to be. It's the essence of living in a herd. The only reason they've survived fifty-million years...is because they have learned how to get along with and depend on one another.
>
> (Rashid, 2004: 81)

In addition to these finely tuned skills in reading body language, horses also need a wise leader in order to survive and they are constantly on the look-out for one in whichever situation they find themselves, whether that is with people or other horses. They will seek a leader in whom they feel secure and can trust; one who is calm and dependable and can keep them safe (Irwin, 2005; Rashid, 2004; Rees, 1984).

A term that has been adopted widely in the EAT/L literature and community is 'mirroring', although this may be better described as 'reacting' (Rees, 2009/2010). Because horses are so effective at reading a person's body language and emotions, it is suggested that they offer a mirror to our own behaviour because they provide immediate and accurate feedback (Ewing *et al.*, 2007; Karol, 2007 ; Klontz *et al.*, 2007; McCormick & McCormick, 1997). If a person is displaying behaviour that the horse feels could be dangerous to its well-being, such as being anxious, agitated or aggressive, it may respond either by acting in a similar way or by simply distancing itself from the person/behaviour by leaving, depending on the circumstances. Conversely, if a person behaves in a calm, confident, consistent and kind manner towards the horse it is much more likely to respond in a calm and cooperative way in return (Mistral, 2007; Rashid, 2004; Rees, 1984; Roberts, 2000). EAT/L can therefore offer a useful and non-confrontational way for a therapist to explore a client's behaviour because

> Horses...give accurate and unbiased feedback, mirroring both the physical and emotional states of the participant during exercises, providing clients with an opportunity to raise their awareness and to practice congruence between their feelings and behaviours.
>
> (Klontz *et al.*, 2007: 259)

However, many factors, such as the age, experience and nature of the horse, together with their environment and many other conditions will have an influence on how the horse responds to a human handler. A better analogy than mirroring might be to think of the horse as providing a useful indicator of our own behaviour because they will certainly react to how they are handled. Research is currently being undertaken which claims that a horse's heart rate quickly corresponds to that of the person handling them and that this could provide useful additional information (Gehrke, 2006; Mistral, 2007).

Other therapists have offered the further dimension that the horse can be a metaphor for a client's behaviour and emotions, drawing on psychological theories of projection and transference (Gordon, 1978; Kohanov, 2001, 2005; Klontz *et al.*, 2007; McCormick & McCormick, 1997). Additionally, it is suggested that clients can relate to a horse's natural desire of wanting to flee when threatened or frightened (Klontz *et al.*, 2007), and that participants seem often to be drawn to particular horses that have similar personalities, behaviours or histories to themselves, opening up further avenues of psychological exploration and self-awareness (Burgon, 2003; Cushing & Williams, 1995; Kohanov, 2001; McCormick & McCormick, 1997; Rector, 2005).

Horses and the natural environment

An assumption of The Yard was that, in addition to the learning and therapeutic gain from being with horses, being outdoors in the natural environment was beneficial to the young people who attended on a number of levels. These included the obvious physical benefits from the exercise involved in working with and caring for horses, together with the mental health advantages that have been attributed to exercise (Halliwell, 2005; MIND, 2007). In addition to these benefits are the less tangible effects of being in nature in perhaps enabling the young people to relax, to be more motivated to learn and engage, and to get away from their problems by being outside in the open, in the natural environment. These assumptions were largely supported in the research study, and the contribution of the horses in relation to these themes is explored in Chapter 6 where the data that emerged about the perceived benefits that the young people gained from being outdoors in nature from both their own perspectives and those of the adults involved in their care are described.

Nature therapy and biophilia theory

Biophilia theory argues that humans have an innate affinity for the natural world, a need which is probably biologically based and which is integral to our development as healthy individuals (Kellert & Wilson, 1993; Nebbe, 2000; Roszak, Gomes & Kanner, 1995; Wilson, 1984).

28 *Equine-Assisted Therapy and Learning with At-Risk Young People*

More recent supporters of biophilia claim that there is now 'a decade of research that reveals how strongly and positively people respond to open, grassy landscapes, scattered stands of trees, meadows, water, winding trails, and elevated views' (Louv, 2008: 43). These authors suggest that this affinity with the natural environment is linked to our evolutionary success as a species; being attracted to resource-rich landscapes and being interested in nature and animals has developed from our hundreds of thousands of years living as hunter-gatherers (Kellert & Wilson, 1993; Wilson, 1984).

Having 'unity with nature' (Levinson, 1972) has been suggested as having therapeutic value. Nebbe terms this 'nature therapy' and states that 'People evolved with the earth, with and within the natural environment', and therefore we are all part of the larger biosphere with which we have a natural bond (Nebbe, 2000: 391). The author gives examples of how we try to facilitate this basic need to connect with the natural environment by camping, hunting, gardening and other such outdoor activities. Katcher and Wilkins in their study on nature therapy in education support a sometimes controversial biological link for a human-nature need, stating that

> The evolutionary development of the human brain was shaped by the necessity to forage and hunt. As a by-product of this necessity, humans have an innate tendency to pay attention to animals and the natural surroundings.
>
> (Katcher & Wilkins, 2000: 153)

These theories are aligned to the Gaia hypothesis and systems theory – that everything is interrelated (Lovelock, 1979). Because we are connected to nature and have a natural affiliation with it, supporters of this hypothesis argue that the natural environment can be a therapeutic setting to aid healing and bring people back in touch with a lost connection to nature and to ourselves (Katcher & Wilkins, 2000; Linden & Grute, 2002; Nebbe, 2000; Roszak, Gomes & Kanner, 1995). It is argued that modern industrialised society's obsession with materialism and technology has resulted in an alienation from nature which is leading to devastating consequences for both children and young people's development, and the environment in general (Louv, 2008). A study in the USA claims that during 1997–2003 there was a 50% decline in the proportion of young people aged between 9 and 12 who spent time outdoors hiking, fishing, gardening and in beach play (Louv, 2008). In the UK a report in 2009 found that fewer than 10% of children reported playing in natural places, whereas 40% of adults reported playing in such places when they were young (England Marketing, 2009). These figures are blamed on a combination of factors, ranging from a rise in technology, computer games and a consumerist culture (Louv, 2008), parental fear of allowing children to play outdoors unsupervised (Ball, Gill & Spiegal, 2008; Durr, 2008), together with busier lifestyles and a reduction in unstructured outdoor play space (Burdette & Whitaker, 2005).

Exercise and health

There is a growing concern that rising obesity levels and mental health problems are linked in part to a lack of exercise, which can be the consequence of our more materialistic, sedentary, unhealthy and technological lifestyle (British Medical Association, 2005; Halliwell, 2005; Louv, 2008; Munoz, 2009). These claims are supported by research claiming that contact with nature and the outdoors can bring both physical and mental health benefits (Ebberling, Pawlak & Ludwig, 2002; Halliwell, 2005; McCurdy et al., 2010; MIND, 2007), is useful in children's development, concentration and ability to learn (Kellert, 2002; Taylor, Kuo & Sullivan, 2001; Wells, 2000), and leads to stress reduction (Korpela et al., 2001). Conversely, it is argued that not having enough time outside is leading to vitamin D deficiency in children (Misra et al., 2008). As a result of these claims, government policy is beginning to support 'green' exercise and to promote contact with nature and the outdoors (BTVC, 2009; Halliwell, 2005; Munoz, 2009). Subsequently there has been a growth in horticulture and nature therapy programmes (Linden & Grute, 2002; Sempik, Aldridge & Becker, 2005), together with referrals by GPs to 'green gyms' and exercise initiatives, such as walking exercise (BTCV, 2009; Halliwell, 2005; Priest, 2007). Alongside this has been a modest rise in the growth of AAT and EAT/L in the UK in prisons, schools, alcohol and drug rehabilitation centres, special education programmes, mental health settings, and residential centres for people with physical and learning disabilities (Bexson, 2008; Donaghy, 2006; EST, 2010; FCRT, 2010; HM Prison Service, 2004; LEAP, 2010; PAT, 2010; SCAS, 2010), although these are currently still far more established in the USA (Chandler, 2005; Mallon, 1994).

A report by Peacock, Hine and Pretty (2007) concentrated on 'green exercise' or 'ecotherapy' for mental health. Research was carried out by the University of Essex with the conclusion that 'these studies confirm that participating in green exercise activities provides substantial benefits for health and wellbeing' and, furthermore, that 'green exercise has particular benefits for people experiencing mental distress' (Peacock, Hine & Pretty, 2007: 1). In addition the authors report that green exercise directly benefits mental health (lowering stress and boosting self-esteem), improves physical health (lowering blood pressure and helping to tackle obesity), provides a sense of meaning and purpose, and 'helps to develop skills and form social connections' (Peacock, Hine & Pretty, 2007: 4). Another study by Sempik, Aldridge and Becker (2005) on the value of social and therapeutic horticulture (STH) focused on the perceived benefits that clients of the five projects that they studied claimed that they received from participating in STH. These included being outside in the fresh air, connectedness with nature, safety, healing, peacefulness and relaxation, distraction (from problems), freedom and escape. Participants spoke of how being outside in the natural environment in all weathers made them feel 'free' and 'peaceful', and how it was good to 'get back to nature' (Sempik, Aldridge & Becker, 2005). The authors make the suggestion that the emotional bond that the participants made with their environment 'could be taken to have a spiritual meaning', in terms of spirituality meaning a connectedness with an entity larger than ourselves (Sempik, Aldridge & Becker, 2005: 49). They conclude that 'the outdoor, natural environment provides the emotional or psychological backdrop for the many activities and processes of STH' (Sempik, Aldridge & Becker, 2005: 52).

Within this literature on the benefits of green exercise and health, 'children have been identified as one of the key social groups to examine' (Munoz, 2009: 9). Reasons for this range from concerns over childhood obesity rates to the importance of unstructured or 'free' play for learning the management of risk, amongst other developmental milestones (Ball, Gill & Spiegal, 2008; British Medical Association, 2005; Ebberling, Pawlak & Ludwig, 2002; Staempflif, 2009). It is claimed that being in nature can alleviate attention deficit hyperactivity disorder (ADHD) symptoms in young people (Taylor, Kuo & Sullivan, 2001), and create a sense of well-being and attachment to place, which can alleviate stress and release tension built up in other areas of their lives (Korpela *et al.*, 2001).

Nature as therapeutic space and for getting away from problems

Berger and McLeod (2006) propose that being in the natural environment can serve as a powerful space for therapy to take place. They suggest that it can 'flatten hierarchies' because

Nature is a live and dynamic environment that is not under the control or ownership of either therapist or client. It is an open and independent space, which has existed before their arrival in it and will remain long after they depart from it.

(Berger & McLeod, 2006: 82)

What Berger and McLeod refer to as 'sacred spaces' has parallels with what Thomas and Thompson (2004) term 'special' or 'secret' spaces; places in nature which are important for children in terms of attachment to place and identity formation. Another area which perhaps has resonance with EAT/L is the idea that each part of nature – the landscape, the elements, the weather, animals – can 'trigger parallel psychological and spiritual quests that can open a channel for mind-body work' and that nature can act as 'a bridge between people' (Berger & McLeod, 2006: 87). The authors also draw from principles of the deep ecology movement, which claims that modern, capitalist society, which concentrates on the individual rather than the collective, is resulting in catastrophe for both humans and the environment with which we are intrinsically linked (Devall & Sessions, 1985; Roszak, Gomes & Kanner, 1995). It is suggested that by reconnecting with nature we can somehow gain a stronger connection to the universe, and this can help to take people outside of their own problems and issues. Kaplan (1995) and Korpela *et al.* (2001) call this 'attention restoration theory' and propose that because we have a fascination with the natural environment and it takes us away from our usual routines, it can help with 'clearing away random thoughts and recovering directed attention' in a way that creates 'restoration' (Korpela *et al.*, 2001: 576). These authors draw on studies with adolescents where they describe their experiences in their favourite places in nature as places to calm down, clear their minds and face troublesome matters (Korpela *et al.*, 2001). A study by Wells and Evans (2003) found that a rural environment, or even just having a view of nature, helped to reduce anxiety among highly stressed children.

These are themes that emerged in the research study and have parallels with some of the experiences obtained through working with horses, with examples of how the young people and adults perceived the role of the natural environment provided in Chapter 6.

Learning and taking risks outdoors with horses

Working with horses involves an element of risk. This is on many levels: from their sheer size and power to their inherent 'wildness' and how, ultimately, they will act on instinct if frightened, despite

32   *Equine-Assisted Therapy and Learning with At-Risk Young People*

millennia of domestication. The young people who attended The Yard learnt how the horse is still closely aligned and attuned to its natural environment – how it is highly influenced by the weather, for example. Many of the young people lacked confidence around the horses when they first started attending TH but grew in confidence once they gained knowledge and skills.

The importance of risk-taking for children and young people's development is well documented, with government policy beginning to recognise this (Ball, Gill & Spiegal, 2008). Many authors write about how participating in outdoor activities which incorporate an element of challenge, such as outward bound courses, ropes courses, a surf programme for at-risk young people, and Green Gyms, for example, can bring both physical and mental health benefits (BTCV, 2009; Carns, Carns & Holland, 2001; Morgan, 2010; Staempfli, 2009; Ungar, Dumond & McDonald, 2005). In their literature review of adventure-based interventions, Moote and Woodarski (1997) propose that these types of therapeutic intervention may be particularly appropriate and useful during adolescence. They suggest that this can be a critical time for at-risk young people who may be drawn to unhealthy and risk-taking behaviours and activities during this period of their lives for a range of reasons related to their psychosocial and childhood experiences and circumstances.

It has been argued that it is especially important for children to have opportunities to engage with their natural environment because it has been shown to 'have a profound effect on children's cognitive functioning' (Wells, 2000: 790). In addition it is claimed that time spent in nature can help with classroom behaviour and academic achievement (Blair, 2009), and may be 'particularly important for young people from deprived backgrounds' (Munoz, 2009: 6) as this population is overrepresented in poor, urban neighbourhoods with limited access to the natural environment (Wells, 2000). A further claim proposed by Nebbe (2000) is that new, more positive ways of behaving can be learnt by being in nature. She suggests that 'the natural environment is a behavior setting that evokes coping rather than defensive behaviors', going on to add that examples of coping behaviours are 'self-sufficiency, risk taking, initiative, and cooperation' (Nebbe, 2000: 394). These claims are supported by Burdette and Whitaker (2005), who add that the opportunity for free play in the natural environment, in particular, provides the opportunity for young children to learn problem-solving skills, and what they term 'attention, affiliation and affect' (Burdette & Whitaker, 2005). It was found that many of the young people who attended The Yard appeared to be more motivated to learn in the context of the outdoor environment with the horse as the focus or the 'backdrop to activities', as will be seen later. The TH programme at The Yard sought to provide engaging, motivating activities through working and learning about horses in the natural environment, and in Chapter 6 some of these in relation to themes of the natural environment, ecotherapy and exercise are explored.

Conclusion

Literature in the field of EAT/L has so far followed two quite disparate paths. On the one hand is the anecdotal, and often spiritual, first-person accounts of how horses can heal individuals (Llewellyn, 2007; Isaacson, 2009; Symington, 2012); on the other is the demand for more 'evidence-based' empirical studies to support the work of practitioners in order to secure its place as a credible field and therefore obtain funding to practice EAT/L, which are costly interventions to provide (Bachi, 2012; Frewin & Gardiner, 2007; Gehrke, 2006; Vidrine, Owen-Smith & Faulkner, 2002). Consequently it has been seen that much of the research has tended to seek to employ psychological testing techniques to measure performance indicators before and after participating in EAT/L sessions (Ewing *et al.*, 2007; Kaiser *et al.*, 2004; Kemp *et al.*, 2013; Klontz *et al.*, 2007; Schultz, Remick-Barlow & Robbins, 2007; Trotter *et al.*, 2008), although it is acknowledged that these studies would not be considered as the most robust in terms of the gold standard of randomised controlled trials (Bachi, 2012; Frewin & Gardiner, 2007).

This book provides a contribution to the gap in the research looking at the process of TH, concentrating not so much on the effects of TH on the young people who attended The Yard but on what was going on between the young people and horses. In addition to my own observations and reflections as a practitioner-researcher, I was interested in hearing in the participants' own voices what was happening for them; what horses they were drawn to want to work with and what it was about those particular horses. What themes emerged, did the horses invoke any particular feelings or was the environment a factor? Did the particular style of natural horsemanship practiced at The Yard have any bearing on how the participants felt and the learning/therapeutic environment at the research site?

It is hoped that by the participants being able to provide their own personal accounts of their experiences of TH, in addition to my own observations and those of some of the adults involved in the young people's care, this provides an additional dimension and richness to the growing body of research within the field of EAT/L. Finally, it is hoped that by the participants being involved in the research process, this would also be of benefit to them in terms of their own learning, as well as being of interest to policy-makers who, ultimately, will influence the direction of EAT/L.

2
Young People Considered 'At Risk', and Overview of 'The Yard' and Methods Used

Risk and resilience

The majority of the young people who attended The Yard had been referred due to being understood as being 'at risk' as a consequence of various psychosocial disadvantages. This concept of 'at risk' has arisen from the research literature, which suggests that exposure to certain risk factors in childhood has a high likelihood in correlating to negative outcomes in later life (Bannister, 1998; Hughes, 2004; Jackson & McParlin, 2006; Masten, Best & Garmezy, 1990; Rutter, 1985; Stein, 2006). In addition, over the last 30 years or so a substantial body of research has concentrated on looking at why certain individuals appear to overcome adversity and remain resilient despite experiencing these childhood risks. Resilience is suggested by Masten, Best and Garmezy, as being, 'the process of, capacity for, or outcome of successful adaptation despite challenging or threatening circumstances' (1990: 426). In contrast with the literature that sees resilience as related to individual personality traits, some authors have argued that it is important to recognise resilience as a 'dynamic state' which understands that risk varies across the life span (Cicchetti & Rogosch, 1997; Luther, Cicchetti & Becker, 2000; Place *et al.*, 2002). These authors suggest that successful resilience is where protective factors can adapt to changing risk and, having experienced successful coping strategies in the past, the ability to cope in the future is strengthened. The work of Ungar is also relevant, for he raises the concern about 'who defines good outcomes and appropriate developmental pathways', and he argues that the important factor is how young people self-define their own health (2004: 24).

Risk factors

Understanding of risk factors varies, depending on perspective, from an emphasis on wider structural understanding of the risks associated with socioeconomic status, class and environment (Bronfenbrenner, 1986; Howard, Dryden & Johnson, 1999; Werner & Smith, 1992) to prioritising the role of neurobiology in individual differences in stress, trauma and risk research (Gunner *et al.*, 2006; Gunner & Quevedo, 2007). In addition there are the predominant behavioural and developmental perspectives on risk and resilience that emerged alongside developmental psychopathology during the 1960s and 1970s (Cicchetti & Rogosch, 1997; Lazarus, 1966; Lazarus & Folkman, 1984; Masten & Obradovic, 2006; Masten, Best & Garmezy, 1990; Skinner & Zimmer-Gembeck, 2007; Werner, 1993). Whilst stress is now acknowledged within the medical professions as having an effect on physical health (Barr, Boyce & Zeltzer, 1996; Halliwell, 2005; Levine, 2005), this book is primarily interested in psychosocial understandings of stress, trauma and resilience in young people.

Risk factors can be experienced both directly and indirectly by an individual and are interrelated. Which social class a child is born into can be considered a distal risk factor as it is not experienced directly. However, class can have a bearing on proximal risk factors, which is where risks are experienced directly – for example, where socioeconomic status results in poor nutrition and housing, which can impact on health and development (Masten, Best & Garmezy, 1990). Being born with a disability can limit an individual's opportunities, but this may be experienced differently depending on socioeconomic status, family and community attitudes, and available resources. It has been suggested that experiencing a single risk factor is not as damaging as exposure to multiple or cumulative risk factors (Garmezy & Masten, 1994; Rutter, 1979, 1999; Werner & Smith, 1992). Looking at the impact of psychosocial risk factors, such as parental discord, the mother experiencing mental health problems, low socioeconomic status, household overcrowding, family criminality and being in care, Rutter (1979) found that whilst experiencing one risk factor alone did not have a significant impact on successful childhood development, measured by prevalence of psychiatric disorder, experiencing two or more of the stressors decreased the opportunity for successful childhood development. Conversely, a criticism directed at much of the literature concerned with risk and resilience is that it has followed a medically based psychological model, concentrating on looking for universal patterns and laws in how individuals respond to stress

(Lazarus, 1993), which does not allow for individual difference, and follows a normative understanding of development (Clarke, Hahn & Hoggett, 2008; Glantz & Johnson, 1999). More recent research on factors related to risk and resilience has evolved from this psychological, medically based, linear model of a stress stimuli and response understanding, to a more transactional, ecological model. This latter perspective recognises that there is a complex relationship between an individual, their environment and the interactions between them to exposure to, and the effects of, risk factors. It argues that an approach is needed which explores the interaction between genetic, biological, psychological, cultural, environmental and socioeconomic factors within a developmental framework in the study of risk and resilience theory (Bronfenbrenner, 1986; Haggerty *et al.*, 1996; Howard, Dryden & Johnson, 1999; Place *et al.*, 2002; Skinner & Zimmer-Gembeck, 2007).

Despite the various differing perspectives on risk outlined, there does appear to be general agreement that exposure to early stressful and traumatic events – physical, sexual or psychological abuse, neglect and dysfunctional parenting, and parental drug and/or alcohol abuse, for example – can result in subsequent 'negative life outcomes' (Bannister, 1998; Egeland *et al.*, 2002; Gunnar *et al.*, 2006; Howard & Johnson, 2000; Hughes, 2004; Jackson & McParlin, 2006; Masten, Best & Garmezy, 1990). Emotional and behavioural symptoms in children and young people which have been attributed to exposure to these risk factors range from hypervigilance, hyperactivity and extreme sensitivity, abrupt and extreme mood swings, lack of self-worth, depression, avoidance and addictive behaviours, to self-harming and the inability to form meaningful relationships (Egeland *et al.*, 2002; Gunnar *et al.*, 2006; Levine, 1997, 2005; Masten, Best & Garmezy, 1990). Levine (2005) goes on to suggest that illnesses such as skin and digestive disorders, an impaired immune system and many other physical conditions can manifest as a consequence of early abusive and traumatic experiences. The 'negative life outcomes' in adulthood that are attributed to risk factors in childhood range from drug and alcohol abuse, mental health problems, criminality and unemployment to failed relationships, ill health and early death (Born, Chevalier & Humblet, 1997; Haggerty *et al.*,1996; Howard, Dryden & Johnson, 1999; Howard & Johnson, 2000; Masten & Obradovic, 2006; Stein, 2006). Many of the young people who attended The Yard had been taken into foster care due to assessed serious risks to their health and well-being. They often came from deprived and impoverished backgrounds and with many of the behaviours listed above routinely reported in their referral forms. Preventative therapeutic

38 *Equine-Assisted Therapy and Learning with At-Risk Young People*

interventions, such as EAT/L and TH offered at The Yard, are concerned with attempting to provide these young people with some of the resilience and protective factors necessary to avoid possible negative life outcomes.

Coping and resilience

Whilst it is understood that risk factors can lead to negative life outcomes there is also recognition that individuals have different responses to and outcomes following negative experiences during their life span. Traumatic life events, such as loss of a parent, loss of a spouse and divorce, for instance, result in stress for most individuals concerned in the same way that certain viruses are likely to produce physical illness in the majority of people but the extent to which people succumb psychologically varies enormously. In the 1960s and 1970s, largely as a result of research on the effects of stress and trauma on the performance of soldiers in World War II and the Korean War, certain researchers became aware that some individuals have different responses and levels of stress to similar events which led to an interest in the area of coping and resilience factors in relation to individual outcomes of exposure to stress (Haggerty *et al.*, 1996; Lazarus & Folkman, 1984; Lazarus, 1993; Levinson *et al.*, 1978, Rutter, 1979, 2006). Why some children and young people develop successfully, despite being exposed to the risk factors outlined previously, has been the subject of research which has labelled these children and young people as 'resilient' or 'stress-resistant' (Garmezy, 1996; Haggerty *et al.*, 1996; Howard, Dryden & Johnson, 1999; Lazarus, 1993). The initial phase of research in this field was often concerned with seeking to discover universal characteristics of resilience related to nature/nurture theories (i.e. negative early experiences leading to psychiatric illness) – influenced by the emergence of Bowlby's (1984) attachment theory – together with genetic reasons for poor developmental outcomes (Barr, Boyce & Zeltzer, 1996; Rutter, 2006; Scarr & McCartney, 1983). Along with a rise in interest in behavioural psychology, research then began to introduce the concept of individual appraisal as a significant factor in resiliency:

> psychological stress, therefore, is a particular relationship between the person and the environment that is appraised by the person as taxing or exceeding his or her resources and endangering his or her well-being.
>
> (Lazarus & Folkman, 1984: 21)

Later, third-phase research began to focus on the link between person and place, and how each impacts on the other. Importantly it also began to recognise that individual resilience varies according to gender, age and circumstances (Compas, Hinden & Gerhardt, 1995; Hampel & Petermann, 2006; Masten, Best & Garmezy, 1990; Rutter, 1985), and that socioeconomic, environmental, structural and cultural factors are involved (Clarke, Hahn & Hoggett, 2008; Howard & Johnson, 2000; Howard, Dryden & Johnson, 1999; Nettles & Pleck, 1996).

Protective factors and mechanisms involved in resilience – internal and external factors

A widely adopted definition of protective factors employed in the literature states that

> Protective factors refer to influences that modify, ameliorate, or alter a person's response to some environmental hazard that predisposes to a maladaptive outcome.
>
> (Rutter, 1985: 600)

It is proposed that there are several variables associated with resilience, which include internal and external factors, such as particular family conditions, community support systems and interventions, and individual personal characteristics. The particular personal strengths and characteristics of an individual are considered to be internal assets, whilst external protective factors are those 'external strengths occurring within the individual's social context' (Howard & Johnson, 2000: 321). Some of the internal characteristics of resilient children are suggested to be personal qualities, such as cognitive skills, an ability to reflect, social competence, a sense of purpose and future, and mastery, autonomy and self-efficacy (Born, Chevalier & Humblet, 1997; Cicchetti & Rogosch, 1997; Howard & Johnson, 2000; Masten, Best & Garmezy, 1990). The presence of a caring adult and quality of childhood care together with family cohesion are family variables (Kraemer *et al.*, 2001; Rutter, 1999; Ungar, 2004), with external protective factors being the level of community support, such as a positive school, other individuals or organisations that reward competency and determination, so providing motivation and support for the young person (Gilligan, 2004; Wolkow & Ferguson, 2001). In a similar way to the idea suggested that exposure to multiple risk factors is more damaging than experiencing a single risk factor (Garmezy & Masten, 1994; Rutter, 1979, 1999;

40 *Equine-Assisted Therapy and Learning with At-Risk Young People*

Werner & Smith, 1992), it is argued that 'the more protective factors that are present in a child's life, the more likely they are to display resilience' (Howard, Dryden & Johnson, 1999: 310).

A cognitive-behavioural approach has been concerned primarily with those personal psychological characteristics which appear to offer an individual resilience, whereas other authors have taken a wider socioeconomic approach based on Bronfenbrenner's (1979) ecological systems theory (Howard, Dryden & Johnson, 1999), together with developmental models which recognise the effects of different stages of child development on individual responses to stress (Hampel & Petermann, 2006; Masten, Best & Garmezy, 1990; Masten & Obradovic, 2006). Depending on perspective and practice orientation, interventions such as EAT/L and TH claim to provide responses and processes which seek to address both internal and external factors, working within developmental models, life-course approaches, and through building skill base and community factors (Lentini & Knox, 2009; Smith-Osborne & Selby, 2010).

Developmental, life-span and ecological models

A development model explores how vulnerability to stress and risk can vary depending on different life stages and how the interactions between age, gender and culture are important in an understanding of development and resilience (Compas, Hinden, & Gerhardt, 1995; Haggerty *et al.*, 1996; Hampel & Petermann, 2006; Masten & Obradovic, 2006; Skinner & Zimmer-Gembeck, 2007). As all of the participants in the research study were adolescents (apart from Lucy at the time of the study, but she had been attending The Yard since her teens), the research study concentrates on looking at this stage of development on factors which may be of relevance, and in how participating in TH may assist in building resilience for this age and risk group.

Interest in adolescent development has a long history and has experienced rapid growth in recent times, influenced by research asserting that adolescent morbidity rates have increased in recent decades in contrast with a decline in morbidity rates for most other age groups (Compas, Hinden & Gerhardt, 1995). Alongside this statistic is concern for adolescents for a host of other reasons amid alarmist claims about increased rates of depression, offending behaviour, unplanned pregnancy, substance abuse and so on (Born, Chevalier & Humblet, 1997; Compas, Hinden & Gerhardt, 1995; Haggerty *et al.*, 1996; Wolkow & Ferguson, 2001). The literature on adolescent development is of importance in the context of this study as the majority of the young people who attended The Yard experienced higher risks than their peers, with the possible additional negative psychosocial consequences and outcomes (Jackson & McParlin, 2006; Jackson & Simon, 2005). It puts in context how much these young people have to contend with in addition to the normal process of adolescence, and of the importance of providing opportunities to participate in interventions concerned with building resilience, such as EAT/L, for these groups of excluded and at-risk young people.

Gender differences

Much of the literature has concerned itself with gender differences in relation to risk and resilience, and with developmental stages and emotional and behavioural problems. For example, it is suggested that girls are more likely to experience 'internalizing disorders' and boys display 'externalizing disorders', and that both genders show increased 'maladaptive coping strategies and externalizing problems' as they reach adolescence (Hampel & Petermann, 2006: 409). Masten, Best and Garmezy propose that,

> resilience is related to sex and to development. The vulnerabilities and advantages of one sex over the other may shift with developmentally related changes in cognition, emotion and the social environment. They also may vary with the cultural context.
>
> (1990: 438)

Biological explanations include exploration of the link between hormonal changes, and behaviours and risk (Compas, Hinden & Gerhardt, 1995). However, interest in the perceived increasingly aggressive and antisocial behaviour of boys reaching adolescence, largely accepted to be influenced by an increase in testosterone during puberty, can also be understood in cultural terms of Western norms and values. There are expectations and demands placed on boys to be 'masculine', of controlling emotions by not crying ('big boys don't cry') and being more assertive than girls, for example (Compas, Hinden & Gerhardt, 1995; Compas & Hammen, 1996; Hampel & Petermann, 2006; Werner & Smith, 1992). Whilst psychoanalytic theory would look at the psychological impact of these conflicting messages (Erikson, 1995; Klein, 1975), Moffitt suggests that 'adolescent delinquency is common and prevalent for boys and may even be described as normative and adjustive in that it serves a developmental function by expressing autonomy' (1993: 274). Other authors have proposed that differences can be traced to pre-birth influences, with risk factors being greater for boys due to their higher rates of pregnancy and birth complications. A result of this is that these babies may be 'more irritable', with the possible consequence of difficult mother–baby attachments and relationships, and the impact of this on later emotional and behavioural problems (Werner, 1993; Werner & Smith, 1992). Werner suggests that qualities which promote resilience are identifiable at an early age, with the resilient adults having been described in infancy as 'good natured' and 'affectionate', and 'having fewer eating and sleeping habits that were distressing to their parents than did the infants who later developed serious learning or behaviour problems' (1993: 504). There are obviously a multitude of proposals put forward as to why gender differences exist, but, whilst it was the case that most of the young people referred to The Yard by the foster care agency and the youth offending team were boys, for this study it was an anomaly that the majority of participants were girls.

In their study with young offenders, Born, Chevalier and Humblet found clear differences between genders, after accounting for statistical difference, stating that 'girls are much more resilient to boys' in having favourable outcomes from incarceration (1997: 685). Unfortunately they do not enter into any discussion about the reasons for the gender differences, concentrating on a generic suggestion that the important factors in resilience for this study group were personal, individual factors, such as self-image, attachment to positive others and feelings of guilt, influenced by family environment. Compas, Hinden and Gerhardt (1995) do engage with the literature on gender disparities, suggesting that there are biological factors at play in links between hormones, mood and behaviour, and problems such as depression and aggression. However, reviewing a study which found that girls whose mothers were ill experienced higher levels of stress and depression than their male counterparts, they conclude that environmental, social and cultural factors are also important, explaining this higher level of stress as due to the increased responsibilities and duties demanded of girls due to their gender (Compas, Hinden & Gerhardt, 1995). In their influential life-span study on the Hawaiian Island of Kauai, Werner and Smith (1992) suggest that protective factors differ for males and females, with internal factors such as self-esteem and cognitive skills having a greater impact on positive outcomes for females, and outside influences of support being more important for males. They go on to propose that different variables take on different levels of importance during the various stages of development, suggesting that individual factors such as temperament and health are the most important factors in early childhood, in middle childhood ecological variables such as family constitution are crucial, and in adolescence it is peer relationships and interpersonal factors such as self-esteem which have the greatest impact on successful developmental outcomes (Werner & Smith, 1992; Werner, 1993).

Adolescence and risk

In addition to interest in gender differences, the research literature notes that adolescence is a critical stage for all young people, regardless of sex, bringing with it hormonal changes together with increasing independence, the possible risk-taking activities associated with this, and the growing importance of relationships with peer groups (Compas, Hinden & Gerhardt, 1995; Erikson, 1995; Hampel & Petermann, 2006; Jackson, Born & Jacob, 1997). For at-risk young people, adolescence can therefore bring with it increased hazards as they may not be equipped with the protective factors of their peers. Strong identification with a significant peer group is proposed to be one of the protective factors in adolescence and it is proposed that

> High identification leads to greater emotional and informative support from the group, whilst low identification is associated with a more marginal status. The latter group tends to have lower levels of self-esteem and cope less effectively.
> (Jackson, Born & Jacob, 1997: 611)

Many of the at-risk young people who attended The Yard had experienced multiple foster care placements, changes of school, and school exclusions, which, together with the other negative experiences of their childhood, had resulted in them often being unable to build successful relationships with family and peer groups. They were therefore perhaps at increased risk of being unable to develop the 'high identification' with peer groups which is suggested as being important. In addition to peer group identification, much of the literature acknowledges that, in the absence of a secure attachment to a loving parent, the next best thing is for the young person to have a 'scaffolding' or network of other positive supportive relationships within the community (Gilligan, 2004). Aspects of this in relation to EAT/L are explored later in this chapter.

Neurobiology of stress and development theories

Other authors take a neurobiological approach to risk and resilience, framed within a child-development perspective and aligned to flight, fight or freeze theories (Bracha, Ralston & Matsukawa, 2004; Cannon, 1929; Gray, 1988). This approach looks at the effect of stress on the neural system and how this can impact on later life responses to stressful events. At a very basic level, when the body anticipates threat, hormones, including adrenalin and cortisol, are released which trigger the individual to react (Seyle, 1950). The individual's decision to fight, flee or freeze as a result of the perceived threat is influenced by a number of developmental and psychosocial factors, but, in the case of a vulnerable child, the freeze, or dissociation, response would often seem to be their main option (Bannister, 1998; Lyons-Ruth, 2003), although increased aggression in other environments can be a consequence (Lyons-Ruth, 2003; Price & Glad, 2003). In addition, it is argued that prolonged exposure to stressful and traumatic events during childhood can impact on emotional and cognitive development due to the negative consequences of long-term raised hormone levels on neural pathways (Gerhardt, 2004; Gunnar et al., 2006; Gunnar & Quevedo, 2007).

Proponents of a neurobiological approach argue that it can explain differences in why some individuals respond to development and social learning-based interventions where others fail to do so; 'psychosocial models often fall short of fully explicating why some individuals respond and others do not' (Gunner et al., 2006: 653). A danger in relying solely on a neurobiological model, however, is that it cannot offer the wider perspective that a psychosocial approach can of the larger environmental and structural factors involved. Furthermore, this model can place an emphasis on traditional parenting and attachment models which may be influenced by cultural and socioeconomic factors (Hollway, 2008). Another criticism of a neurobiological approach is of how effective a model based on studies carried out on rodents and primates can be when translated to human experience, as acknowledged by Gunnar et al.: 'Species differences and variations in the nature of adverse early experiences studied in animals and humans preclude direct translation of the animal studies to the human case' (2006: 672).

There has been recent renewal of interest in a possible link between abusive and violent parenting on the neurobiological pathways of abused children following high-profile cases in the media. 'Baby P' and other similar cases have been seized upon by the popular press

(Foster, 2009; Garner, 2009). In addition, key figures, such as Martin Narey, the director of Barnardo's, have called for more at-risk babies to be removed from 'broken families' before the damage to these children is 'too great to repair' (McVeigh, 2009). These arguments draw on studies from neuroscience and biochemistry to claim that the limbic system and brain patterns of abused and neglected children have been formed in response to cruel and violent treatment and, consequently, they have not been able to develop traits such as empathy and kindness (Foster, 2009; Garner, 2009). In addition, learned behaviour theories suggest that children learn to inflict the violence that has been done to them onto others (Garner, 2009). Clearly these are emotive and political issues; a socioeconomic response would argue that concentrating on neurobiology and blaming individual parents and children avoids the bigger issues of societal inequality and the consequences of prolonged poverty and hopelessness (Hoggett, 2008; Howard & Johnson, 2000; Masten & Obradovic, 2006).

Cognitive-behavioural approach

A cognitive-behavioural approach to resilience and coping draws heavily on a long Western philosophical tradition of 'the idea that how a person *construes* an event shapes the emotional and behavioural response' to it (Lazarus & Folkman, 1984: 24). The work by Lazarus (1966) introduced the concept of how an individual's appraisal of a stressful situation is an important factor in how they respond to threat. An interpretivist perspective similarly follows the standpoint that how something is perceived by an individual is always subjective and 'should be considered in terms of their significance to the individual' (Lazarus & Folkman, 1984: 25). Cognitive-behavioural therapy (CBT) emerged alongside this perspective (Beck, 1976; Beck *et al.*, 1979; Ellis & Greiger, 1977) although it is not without its critics, who argue that it is due to its emphasis on short-term or 'quick fix' treatment, that it is has become popular politically due to it being cheaper to administer than longer-term psychoanalytic and psychotherapeutic approaches (Hoggett, 2008). Cognitive-behavioural approaches to resilience have subsequently concentrated on interventions that are concerned with assisting individuals to find internal resources to modify their behaviour in order to react in a more positive or productive way in the face of adverse external stimuli. It is suggested that certain characteristics may help to provide resilience or protection from risk. These include self-efficacy and a sense of mastery, self-esteem, learned resourcefulness/problem-solving

46 *Equine-Assisted Therapy and Learning with At-Risk Young People*

skills, constructive/positive thinking, social competence, and a sense of purpose and future, although the contexts and environments in which they are located are recognised as interrelated factors (Born, Chevalier & Humblet, 1997; Howard, Dryden & Johnson, 1999; Lazarus, 1993; Masten, Best & Garmezy, 1990; Rutter, 1985).

If a culture of helplessness and hopelessness with no sense of future is considered to be one risk factor to successful development, it is therefore argued that acquiring a sense of mastery and feeling of having some control over one's destiny can promote resilience (Bandura, 1982; Cicchetti & Rogosch, 1997; Katz, 1997; Masten, Best & Garmezy, 1990). Many of the at-risk young people who attended The Yard experienced social exclusion and powerlessness of one form or another. This could be by being excluded from school, or as a consequence of moving between numerous foster placements and therefore being out of the school system. In addition, many displayed emotional and/or behavioural difficulties due to their traumatic past. Together these factors could result in difficulty in forming positive relationships, with further exclusion being a consequence. One result is that many of these young people get left behind in comparison with the developmental and educational milestones of their peers, with all the additional stigmas and effects on self-esteem and confidence that this causes. Another is lack of opportunity, with the consequent limitation on horizons and sense of future: only 6% of care leavers achieve five or more GCSEs at grade C and above, compared with 53% of all children, and only 5% enter higher education (Jackson & McParlin, 2006; Jackson & Simon, 2005). A sense of success, achievement and developing a talent in something significant to the individual is suggested to be key to successful child development (Berman & Davis-Berman, 1995; Carns, Carns & Holland, 2001; Ellis, Braff & Hutchinson, 2001; Katz, 1997). It is therefore especially important for an at-risk young person who may have only felt failure, limited opportunities and negative experiences to have the opportunity to participate in activities which help to develop their strengths and lead to positive successes, a growth in self-confidence, self-esteem and mastery. This activity could then be considered as a protective factor as

> Self-efficacy increases as a result of mastery experiences; in turn, feelings of self-efficacy increase the likelihood of instrumental behaviour. Resilient children may enter a situation more prepared for effective action by virtue of their self-confidence.
>
> (Masten, Best & Garmezy, 1990: 431)

'Acts of helpfulness'

A further proposed element of self-efficacy is the concept of 'required acts of helpfulness' or 'required helpfulness' (Katz, 1997; Howard, Dryden, & Johnson, 1999; Werner, 1993). It is suggested that having an area of responsibility which involves caring for something or someone who is dependent on, and benefits in some way from, that care, *including animals,* can lead to a growth in self-esteem and self-efficacy (Gilligan, 1999; Myers, 2007; Werner, 1993). Acts of helpfulness can lead to the individual learning that they can effect positive change and have something valuable to offer (Howard, Dryden & Johnson, 1999; Katz, 1997; Werner, 1993). In relation to this research study the themes of self-esteem, sense of mastery and self-efficacy would seem to be relevant, in terms of many of the young people who participated in TH reporting feelings of achievement from overcoming challenges, learning a new skill and increased confidence. These were also reported by the adults in the young people's lives who were interviewed. Additionally there are the acts of 'required helpfulness' benefits gained from the caring and responsibility involved in looking after horses.

To date, most interventions aimed at promoting resilience in young people within a behavioural model have concentrated on providing services within the school and statutory health and social care sectors based on modification techniques. These interventions have looked at providing 'preventative packages', such as behaviour management and counselling services, policies to legislate parental responsibility for their children both in and out of school, life-skills programmes, and other interventions aimed at changing behaviours and teaching skills which may promote resilience (Place *et al.,* 2002; Rutter, 1999; Consortium on the School-Based Promotion of Social Competence, 1996). However, this approach has been heavily criticised for failing to take into account cultural differences and values, and for operating from a model which sees the children and families as deficient in some way:

> students labelled by schools as vulnerable or at-risk are often those whose appearance, language, culture, values, home communities, and family structures do not match those of the dominant culture.
> (Howard, Dryden & Johnson, 1999: 308)

In addition, it may be the case that what is considered to be a risk factor in one environment or cultural context, such as aggressive or, alternatively, very passive behaviour, may be a necessary survival

48 *Equine-Assisted Therapy and Learning with At-Risk Young People*

strategy in another (Howard, Dryden & Johnson, 1999; Consortium on the School-Based Promotion of Social Competence, 1996). Rutter warns against a tendency to attempt to draw universal conclusions about what constitutes resilience because 'resilience does not constitute an individual trait or characteristic', arguing that the focus of research should be on protective processes rather than protective factors (1999: 135).

It was understood by The Yard that the young people referred were located within a larger socioeconomic system that labelled them deficient in certain ways. Whilst structural, ideological, criticisms could be levied at The Yard because TH is an intervention that operates within a psychotherapeutic and learning context, as opposed to seeking to challenge the socioeconomic and political system, it was hoped that by working within a 'dynamic state' model and concentrating on the young people's strengths within a person-centred approach, young people could have the opportunity to learn ways to build resilience and obtain protective factors. In turn, it was hoped that this would enable them to make more positive life choices which would be helpful for them within their own personal circumstances. In addition The Yard sought to open up the 'positive opportunities' suggested by Rutter (1999) that can be provided by community interventions into which EAT/L and TH fall.

Community protective factors and EAT/L

Whilst it has been seen that there is a comprehensive body of research interested in exploring the particular personal characteristics, family relationships and social conditions that foster resilience in children and young people, there is less on what are sometimes defined as community factors (Wolkow & Ferguson, 2001). These include community-based interventions that are referred to as therapeutic recreation programmes and adventure or challenge courses/therapy (Berman & Davies-Burman, 1995; Carns, Carns & Holland; 2001; Ewert, McCormick & Voight, 2001; Moote & Wodarski, 1997; Voight, 1988). Despite the relative lack of evidence, many authors argue that community support interventions are valid and important factors in promoting resilience and can be the most effective, as other protective factors, such as family dynamics and personal characteristics, may be more difficult to influence and change (Gilligan, 1999, 2004; Ungar, Dumond & McDonald, 2005; Wolkow & Ferguson, 2001). Rutter talks of the importance of alternative positive experiences for at-risk young people, stating that research has shown 'the substantial benefits that may come from turning point experiences in early adult life that provide a discontinuity with the past and *open up new opportunities*' (1999: 136 added emphasis). It will be seen in Kelly's last day in Chapter 3 how at a time of powerlessness and displacement The Yard gave her the opportunity to re-author herself by opening up new future possibilities for her.

Traditional community-based interventions include participation in after-school centres and clubs, together with some of the alternative therapeutic interventions and outdoor adventure courses outlined above, largely initiated in the USA. The recent growth in research exploring the link between health and the natural environment (Louv, 2008; Sempik, Aldridge & Becker, 2005; Taylor, Kuo & Sulivan, 2001; Wells, 2000) has resulted in it being considered at policy level that outdoor play and experiences are important for both children and adults, with subsequent growth in gardening, forest and nature projects (Munoz, 2009; O'Brien & Murray, 2005; Sustainable Development Commission, 2008). It is claimed that effective community-based interventions can incorporate many of the protective factors raised in the literature as being necessary for successful development and building resilience. These include increasing self-esteem and the self-confidence gained from learning and mastering challenging new skills, together with improving communication, and peer and social skills (Moote & Wodarski, 1997; Place *et al.*, 2002; Sempik, Aldridge & Becker, 2005; Wolkow & Ferguson, 2001). In addition, these interventions can act as a further protective mechanism by providing an alternative to high-risk peer groups and environments (Rutter, 1985, 1999), and the additional physical and mental health benefits of participating in outdoor activities that can be achieved (Carns, Carns & Holland, 2001; Halliwell, 2005; MIND, 2007; Ungar, Dumond & Mcdonald, 2005). Moote and Woodarski (1997) suggest that creative therapeutic interventions may be particularly appropriate and useful during adolescence as this can be a critical time for at-risk young people who may be drawn to unhealthy and risk-taking behaviours and activities.

Community interventions can also provide the opportunity for the at-risk young person to develop the positive relationships necessary for successful development. It is suggested that it is important for children to have warm and non-critical relationships and to be unconditionally accepted in order to develop the sense of self-worth which leads to self-esteem and a positive sense of self (Bowlby, 1984; Karen, 1998; Winnicott, 1965). For young people lacking a positive attachment to a parent it may be that 'Other adults outside the family may also be very

50 *Equine-Assisted Therapy and Learning with At-Risk Young People*

important, for instance teachers, neighbours and mentors' (Gilligan, 2004: 95). Gilligan goes on to add that it is important that these relationships are meaningful and that a 'therapeutic alliance' is formed (Gilligan, 2004). From a relationship of 'unconditional positive regard' (Rogers, 1951) it is hoped that the young person can gain the confidence and motivation to try new, perhaps challenging, activities, leading to new skills and competencies which can lead to a growth in self-esteem.

EAT/L, together with TH practised at The Yard, fit into the categories of therapeutic and recreational activities mentioned previously and, whilst the evidence base is still limited, initial findings support similar benefits and outcomes as those outlined above (Bowers & Macdonald, 2001; Burgon, 2003; Cawley, Cawley & Retter, 1994; Ewing *et al.*, 2007; Kaiser *et al.*, 2006; Lentini & Knox, 2009; Vidrine, Smith & Faulkner, . 2002). In this research study the themes that emerged of increased self-confidence, challenge, empathy and self-efficacy are argued to help to provide participants with some of the protective factors necessary to increase resilience.

Criticisms of resilience theory

It would appear that how an individual appraises and subsequently copes with a stressful situation depends on a whole host of individual biological, cultural, environmental and socioeconomic factors, with a corresponding range of research perspectives involved. As has emerged from the literature, it is generally accepted, across perspectives, that there is a relationship between childhood stresses and later life risk factors, such as mental health problems and risky behaviour, such as drug and alcohol abuse and offending. Young people suffering disadvantage, such as those attending The Yard, have been labelled 'at risk' within the literature and statutory services. A criticism of this approach, stemming as it does from a medical/biological model, is that it sees the individual as the problem, and their behaviour as defective and in need of changing. Behaviour is measured against a 'normative' model with the implied assumption that 'normal' behaviour is desirable (Howard, Dryden & Johnson, 1999; Laing, 1968; Luther, Cicchetti & Becker, 2000; Ungar, 2004). Illich, talking of health in its widest sense, states that culturally we have created a 'vision of health as normality and of sickness as deviance from a theoretical norm' (1975: 89). A cognitive-behavioural approach therefore concentrates on interventions to change the person rather than looking at the structural, political and socioeconomic processes behind these behaviours, whereas it is argued that a more critical perspective is required which engages with the bigger issues of inequalities in society (Bourdieu, 1999; Dominelli, 1997; Fook, 2002 Hoggett, 2008; Illich, 1975). It could be argued that stress and subsequent emotional behavioural difficulties are a 'normal' response to the lie of the 'American Dream' of Western ideology and 'the mistaken view that any and all could succeed were they to work hard' (Haggerty *et al.*, 1996: 13).

Hoggett (2008) following Bourdieu (1999) talks of 'social suffering', referring to the pain, anger and despair brought about by chronic poverty and hopelessness, and how this results in these emotions becoming either internalised leading to depression, alcoholism, addictive behaviours and so on, or externalised as violence against self or others. He also questions how the 'actual subjective experiences and lived lives of those concerned' are lost in the largely quantitative resilience research literature which concentrates on indicators of risk, such as social class, gender and ethnicity, and therefore merely perpetuates positivist norms of devaluing individual experiences (Hoggett, 2008: 80). In order to counter some of the connotations of resilience being an internal personality trait and acknowledge that it is a dynamic state operating within larger contexts, Luther, Cicchetti and Becker caution against referring to 'resiliency' in favour of always employing the term 'resilient' in the literature (2000).

As previously acknowledged, the practice of The Yard, and EAT/L in general, is open to the criticisms above, of working within a deficient model and of being concerned with changing behaviour. Despite commitments to anti-oppressive and person-centred practice it is impossible to deny or avoid the power relations involved, especially highlighted when working with young vulnerable people, who are already doubly oppressed Christensen & James, 2000; Graue & Walsh, 1998; Morrow & Richards, 1996; Robinson & Kellett, 2004). Rather than avoid these issues The Yard chose to work within a relational, person-centred model which sees the person as 'inherently good' (Maslow, 1968; Mearns & Thorne, 2000; Rogers, 1951, 1980) and strived to create a 'safe space' where the young person could explore their issues at their own pace (Axline, 1971; Winnicott, 1965).

A psychosocial perspective would see that a combination of biological, sociological and psychological factors is equally relevant in a study of stress, risk and resilience as summed up by Skinner and Zimmer-Gembeck, who state that

> the development of coping cannot be understood without considering the multiple physiological, emotional, behavioural, attentional,

and interpersonal processes that give rise to it and the larger social ecological contexts within which it unfolds.

(2007: 137)

Framing the 'at-risk' young people who attended The Yard within the larger structural context of their lives, this book looks at how their experiences of participating in TH appeared to provide some of the aspects related to resilience and protective factors. These themes are presented throughout the empirical chapters, but especially in chapters 3 and 4, which explore themes related to self-confidence, self-esteem, opening up of 'positive opportunities', and development of empathy, attachment and trust.

Attachment and trust: Horses and relationship

According to Green and Goldwyn, attachment theory 'is one of the most influential theories in relation to children's social development' (2002: 835). Certainly, within the 'at-risk' population of young people who attended The Yard, 'attachment issues' were often referred to in relation to their case histories and in their assessment forms, perhaps not unsurprisingly considering their fragmented, traumatic and insecure backgrounds. In the field of EAT/L, attachment theory is suggested as a theoretical framework by numerous authors who argue that the horse can provide some of the elements necessary for building positive attachments and relationships where these were previously missing (Bachi, 2013; Bowers & MacDonald, 2001; Gammage, 2008; Karol, 2007; McCormick & McCormick, 1997; Meinersmann, Bradberry & Bright Roberts, 2008; Moreau, 2001; Vidrine, Owen-Smith & Faulkner, 2002).

Attachment theory

Attachment theory has as its foundation the premise that a healthy, secure attachment to a 'warm, sensitive, responsive and dependable' parent or caregiver in childhood is necessary in order to grow into a healthy and successful adult (Karen, 1998: 6). Conversely, experiencing inconsistent parenting, abuse and 'maternal deprivation', as experienced by many of the participants at The Yard, can lead to 'disorganised attachment' and pathology in later life (Boris & Zeanah, 1999; Bowlby, 1984; Crittenden, 1988; Howe, 2005; Hughes, 2004). In his later work, Bowlby updated his theories to include his work on 'developmental pathways', suggesting that we are a product of both nature and nurture, with infants being born with innate characteristics which are then either negatively or positively influenced by their environment (Bowlby, 1988). He proposed that

> Children who have parents who are sensitive and responsive are enabled to develop along a healthy pathway. Those who have insensitive, unresponsive, neglectful, or rejecting parents are likely to develop a deviant pathway which is in some degree incompatible with mental health...
>
> (Bowlby, 1988: 136)

In the late 1930s, Bowlby was at odds with the prominent childhood theories of his time which proposed either that a child's behaviour was predominantly genetically programmed or, conversely, that children were entirely a product of their environment. Practical approaches to child-rearing were generally based on not showing too much affection and installing discipline (Karen, 1998). Within the psychoanalytic tradition of the time, dominated by Freudian and Kleinian theories, the emphasis was very much on the internal world of the infant (Bretherton, 1992). Bowlby, together with his colleague Mary Ainsworth, added a further perspective – that most important was that children received both love and a 'secure base' in order to thrive (Bowlby, 1984, 1988).

Drawing on ethology, psychoanalysis and psychological behavioural theories, Bowlby's early work focused on the distressing effect of maternal separation of the child from their mother. He proposed that psychological attachment to their mother is a 'primary motivational system' in its own right as opposed to the Freudian and Klenian theories of the time which postulated that attachment between mother and infant was based on survival instinct (Bowlby, 1984). The Strange Situation Experiment by Ainsworth sought to give an empirical base to Bowlby's theories, looking at the effects of secure and insecure attachments by noting the reactions and behaviours of year-old babies on their mother's absence and return (Ainsworth & Bell, 1970). The research identified three types of attachment: secure attachment, insecure attachment and non-attachment. Insecure attachment was subdivided into three further categories of anxious resistant or ambivalent attachment, anxious avoidant attachment and disorganised attachment. In a nutshell, the research proposed that securely attached babies will show some distress at their mother's departure but will allow themselves to engage with another carer until their mother's return, at which they greet her with a desire for contact and will allow themselves to be comforted by her. In insecurely attached babies the behaviour varies from clingy behaviour

54 *Equine-Assisted Therapy and Learning with At-Risk Young People*

to higher levels of distress on their mother's absence and return. Contradictory behaviour may also be displayed such as rejecting their mother and other dysfunctional behaviours, ranging from detachment to stereotypical actions such, as head-banging and rocking (Ainsworth & Bell, 1970; Ainsworth & Bowlby, 1991).

The holding environment and EAT/L

A colleague of Bowlby, Winnicott introduced the concept of the 'holding environment', which has many parallels with attachment theory, suggesting that the defenceless infant needs to be given a secure holding base in which to be psychologically able to cope with developing a secure sense of self in the world (Winnicott, 1965). Winnicott proposed that by not having their needs met or by experiencing abusive, inconsistent care, the child learns to develop coping or defence strategies in an attempt to manage their survival. This then can become the foundation for the later development of psychopathology (Winnicott, 1965). Similarly, from an object relations perspective, the failure on the part of the environment to successfully attune to the needs of the infant can create the development of maladaptive behaviours and a 'protective false self' (Frame, 2006; Parish-Plass, 2008).

Winnicott's holding environment has three primary aspects which are holding, handling and object-presenting. Elements of all three are proposed as being relevant in EAT/L (Brooks, 2006; Frame, 2006; Karol, 2007). Only through meeting the necessary physical and psychological needs of the defenceless infant by 'good enough' mothering can the child find their 'true self' (Winnicott, 1965). It is understood in client-centred and psychodynamic therapy that it is through relationships that we can understand who we are, expressed by Winnicott as 'when I look, I am seen, therefore I exist' (1971: 134). Object presentation describes the manner in which the world is introduced to the child by its mother. Through being sensitive and attuned to the infant and encouraging them to discover things for themselves within a supportive and loving relationship, the child is able to gain confidence and trust in themselves and the environment, and consequently gain trust in their own capacity to have their needs met (Winnicott, 1971). Many of the young people who attended The Yard displayed a chronic lack of self-esteem, confidence and difficulty in relationships, and it could be argued that these may be partly attributed to not having received these early positive experiences. It has been suggested that through EAT/L, horses can become the symbolic mother in which the young person reworks earlier childhood experiences that may lead to finding what Winnicott (1965) terms one's 'true self' (Frame, 2006; McCormick & McCormick, 1997). Some of the ways in which it is suggested that horses can help to meet some of the three aspects of the holding environment are through the horse enabling the young person to be able to build a healthy attachment, relationship and trust with a living animal. Where previously physical touch with adults may have been unhealthy and/or conditional, through riding on and having physical contact with a horse, real and metaphorical aspects of being 'held' and being handled are met in a healthy way (Brooks, 2006; Gammage, 2008; McCormick & McCormick, 1997; Yorke, Adams & Cody, 2008). Frank and Frank make the point that most Western psychotherapies avoid bodily contact and touch due to the fear of it being perceived as 'either erotic or aggressive' (1991: 130). A theory proposed is that the weight and size ratio between a child and horse is very similar to that of a baby and its mother, and that this perhaps has significance for 'symbolic mother' and 'holding' theories (Winnicott, 1965), which include the importance of healthy touch (Vidrine, Owen-Smith & Faulkner, 2002).

Talking about animals in general, it is suggested that they can 'satisfy needs for physical, emotional, and social contact without the fear of unwanted or threatening involvement with other human beings' (Becker, cited in Walsh, 2009: 472). Additionally, training and working with horses constantly includes elements of 'object presentation' as the horse, which is inherently fearful due to its instinctive prey status, learns not to be afraid of potentially frightening objects in its environment by taking cues from its handler that there is nothing to fear. It could be that by the young person taking on the role of the 'good enough mother' that they are giving some of these elements back to themselves.

In contrast with the traditional psychoanalytic theories of his time, which stated that psychological pathology was a consequence of regression to fixed stages of childhood development, Bowlby believed that development was more related to the child's experience of their environment. From this perspective the individual could be open to positive change at any stage of life. He believed that therapy could be usefully applied in order to provide similar conditions to the 'secure base' of the mother in order to explore and heal attachment issues. This concept of providing a sense of security to the patient was fundamental to Bowlby: 'unless a therapist can enable his patient to feel some measure of security, therapy cannot even begin' (Bowlby, 1988: 140). In a similar way to Bowlby's belief that the individual is always potentially open to positive change, core process psychotherapy has the primary belief that the 'true self' wishes to be fundamentally healthy. It is simply obscured by the 'protective false self' through defence mechanisms learnt through traumatic and abusive childhood experiences (Mearns & Thorne, 2000; Rogers, 1967). In order for the therapeutic process to be successful, it is understood by client-centred therapists that building a trusting relationship is imperative. This includes the necessity for the therapist to trust in the child's inner capacity to develop and heal (Gammage, 2008). An established principle in psychotherapy is that the strength of attachment between client and therapist, characterised by empathy, acceptance, trust, warmth and collaboration, is the most powerful predictor of positive client outcome (Lambert & Bergin, 1994; Rogers, 1951; Winnicott, 1965; see also Fine, 2000; Schultz, Remick-Barlow & Robbins, 2007; Yorke, Adams & Coady, 2008). In Chapter 4 it is proposed that in some of the encounters between the horses and the young people, the horses provided some of the conditions of the 'secure base', and other aspects that are claimed to be so fundamental in creating a successful therapeutic 'space'.

Criticisms of attachment theory

Bowlby's attachment theory has largely been embraced and accepted into mainstream paediatrics and child development, with his earlier work criticising the routinely established practice of separating babies from their mothers at birth in maternity wards, having successfully resulted in this no longer being normal practice (Holmes, 1993). Nevertheless, Bowlby's theories have also been subject to much criticism, initially by Rutter (1972), who questioned the term 'maternal deprivation' as too simplistic to account for all of the variables involved in the dysfunctional development of children. Later criticism of attachment theory was made by feminist writers who claimed that this emphasis on attachment bonds to the mother risked placing the sole blame on mothers for any dysfunction, rather than the multitude of other influences at play, such as poverty and other sociocultural factors (Birns, 1999). In addition, understanding the mother–child relationship and nuclear family structure as biologically 'natural' can be employed for political purposes because it suits a modern capitalist society very well for the father to be the main breadwinner and the mother to remain at home child-rearing, taking responsibility off the state and placing it on the family (Chodorow, 1978; Leupnitz, 1988). In terms of the relationship between attachment theory and child-protection practice in social work, it is suggested that attachment theory needs to be seen in the context of the mother's wider socioeconomic and life circumstances (Krane *et al.*, 2009). These authors argue that therapeutic engagement needs to be put in place to support the mother to overcome adverse circumstances and possible disorganised attachment patterns of her own (Krane *et al.*, 2009). Other writers have suggested that it is the quality of infant care that is most important, rather than just the mother–infant bond, and that this can be carried out just as well within an extended family network, such as practised in many other cultures (Neckoway, Brownlee & Castellan, 2007). Cleary (1999) makes the point that Bowlby's theories are based on Western, empiricist and, therefore, male-dominated models of understanding. She argues that his theory 'limits our appreciation of the cultural and ancestral influences that organize psychological life' (Cleary, 1999: 41).

Attachment and trust in the context of horses and the therapeutic relationship

A recurring theme in the course of the fieldwork and interviews with participants was of the young people talking about how they could trust the horses, and of how important it was for the horses to trust them. As has been discussed previously, being able to trust is fundamental to the therapeutic relationship, and in order to trust it is necessary to feel safe (Axline, 1971; Maslow, 1968; Rogers, 1951, 1980). Participants spoke of how the horses understood them, how they could trust the horses and also of how the horses needed to feel safe. Additionally, the horses seemed to enable some of the young people who found it difficult to show affection to humans – due to the dysfunctional ways in which they may have experienced affection and physical contact previously – to show affection and empathy towards the horses. They would often hug the horses and talk to them affectionately. A psychotherapeutic interpretation may suggest that beginning to be able to express affection and care towards another is a first step in being able to give it to oneself (Gammage, 2008). It is argued that horses can provide the opportunity to experience a trusting and unconditional relationship together with the healthy tactile element from another living being that are essential for healthy development (McCormick & McCormick, 1997; Yorke, Adams & Coady, 2008). As mentioned earlier, the fact that the body weight ratio between person and horse is similar to the body weight ratio between a mother and infant has been raised as perhaps being significant (Lentini & Knox, 2009; Vidrine, Owen-Smith & Faulkner, 2002). Additionally, the mesmerising rhythm of the horse's gait when being ridden has been cited as perhaps being a contributing factor in encouraging relaxation (Game, 2001) and of 'rhythmic harmonization', a feeling of connection, unity and freedom (Evans & Franklin, 2010). Something that was witnessed time and time again was the young people relaxing and opening up when on the back of the horse, as will be demonstrated from extracts from interviews and fieldnotes in the following chapters. Feeling safe and calm, with the young people often expressing this as the horses' need to feel calm and safe, was another recurring theme to emerge from the research study and also has links to the mindfulness literature.

Mindfulness and links to TH

Mindfulness in its simplest definition is the practice of being aware and attentive in the present moment (Biegel *et al.*, 2009; Germer, Siegel & Fulton, 2005; Kabat-Zinn, 1994). Interventions employing aspects of mindfulness meditation are claimed to help with a variety of psychological problems (Mason & Hargreaves, 2001) and argued to be especially useful with adolescents (Biegel *et al.*, 2009; Wagner, Rathus & Miller, 2006). The Western health model of mindfulness now being adopted in some mainstream health services has evolved from ancient meditative traditions with their roots in Buddhism. In the Mental Health Foundation Mindfulness report, a description incorporating some of the practices involved in achieving mindfulness states:

> Mindfulness is a way of paying attention to the present moment by using meditation, yoga and breathing techniques. It involves consciously bringing awareness to our thoughts and feelings, without making judgements about them. It is a method for observing what is happening right now, in our bodies, minds and the world around us.
>
> (Halliwell, 2010: 16)

It is well documented that traditional psychotherapy and counselling practices may not be the most effective or accessible treatment options for many young people, such as the 'at-risk' population who attended The Yard, because 'some adolescents do not view psychotherapy as a beneficial treatment option' (Biegel *et al.*, 2009: 855). In their study with adolescent psychiatric outpatients, the same authors suggest that mindfulness-based stress reduction (MBSR) techniques may be more appropriate. Goodman (2005) makes the claim that many of the key concepts and practices of mindfulness are especially relevant to children. She states that 'children do not communicate the way adults do. Many of their thoughts and feelings are expressed nonverbally, through play and body gestures', therefore it is important to employ a therapeutic intervention that is receptive to this (Goodman, 2005: 198). Wagner, Rathus and Miller suggest that mindfulness-based exercises 'that engage multiple sensory systems (e.g., listening, tasting, smelling) or that involve movement' are especially effective with adolescents (2006: 186). Because mindfulness-based therapy is based on acceptance and 'few preconceived notions' together with 'psychotherapeutic presence' and the emphasis on being in the present moment, it offers a therapeutic medium which facilitates connecting with children and young people (Goodman, 2005; Wagner, Rathus, & Miller, 2006). In addition, it is suggested that being outside in the natural environment facilitates a mindful state naturally as it sharpens awareness (Coleman, 2006). A number of examples of how working and interacting with horses embodies many of these same qualities found in mindfulness practice and therapies, especially in relation to those that include movement and other sensory experiences, are introduced in Chapter 6.

Some mindfulness practices, such as MBSR and mindfulness-based cognitive therapy (MBCT), are beginning to show promising empirical results and consequently being introduced into mainstream health services (Baer, 2006; Biegel *et al.*, 2009; Brown & Ryan, 2003; Heppner *et al.*, 2008; Kabat-Zinn *et al.*, 1992; Kuyken *et al.*, 2008; Mason & Hargreaves, 2001; Moss, Waugh & Barnes, 2008; Zylowska *et al.*, 2008). Whilst these practices are claiming to show 'how ancient wisdom combined with modern science can improve mental health' (Andrew McCulloch cited in Halliwell, 2010: 2), they possess a fundamental theoretical difference with respect to our traditional goal-driven, results-orientated, Western approaches to the health of the mind and body, of trying to fix the problem. This 'disease-management model' (March, 2009) sees the problem as the 'enemy', a concept not useful in MBSR (which would seem to be the approach most aligned to EAT/L), because it sees that this traps people in patterns of ruminating on problems that they are trying to escape. Mindfulness practices believe that trying to escape and/or shut off or fight negative or unwanted thoughts and feelings can actually cause more stress and tension in the body (Germer, 2005). Instead, MBSR techniques use exercises, stretches, mindful walking and yoga in order to encourage calmness. Body scan exercises, such as those employed in the 'invisible riding' techniques practised at The Yard, focus on bringing awareness to each part of the body in turn, but without judgement. Segal, Williams and Teasdale suggest that

A major aim of the body scan exercise is to bring detailed awareness to each part of the body. It is where participants first learn to keep their attention focused over a sustained period of time, and it also serves to help them develop concentration, calmness, flexibility of attention, and mindfulness.

(2002: 110)

This is also described as 'meditation in motion', an exercise where participants are directed to walk mindfully while keeping a 'soft gaze' (Segal, Williams & Teasdale, 2002). These authors suggest that walking meditation 'can be especially useful to those who feel agitated and unable to settle. The physical sensation of walking, people comment, tends to enable them to feel more "grounded" '(2002: 181). Segal, Williams and Teasdale go on to suggest that when the mind is agitated or a person feels pressured, 'it is easier to be mindful with a practice that involves physical movement than with one that does not' (2002: 181). In Chapter 6, similarities are drawn to the 'invisible riding' exercises practised at The Yard and some of the body-based exercises described in the mindfulness literature, where participants who sometimes found it difficult to concentrate and 'settle' are directed to bring awareness to their bodies in a relaxed, open way without negative judgement. It is shown how these exercises could help participants to achieve a mindful, relaxed and effective relationship with the horse. Another interpretation of mindfulness employs the term 'participant observation' to describe that it 'is not detached witnessing. It is experiencing the mind and body *more* intimately' (Germer, 2005: 9). Shaver *et al.* (2007) state that one of the common goals of different meditation techniques is to achieve 'deep relaxation' together with 'alert attention'. They argue that by experiencing mindfulness on a regular basis, many other benefits can be achieved, such as the *organismic trusting* and *openness to experience* they refer to as related to Rogerian theories of being essential to the 'fully functioning person' (Shaver *et al.*, 2007).

A major difference in mindfulness compared with other therapeutic approaches is that mindfulness is concerned with *being* rather than *doing*. It is not goal-orientated as such, with any benefit being a natural by-product of the practice. It is claimed that by embodying a being mode as opposed to a doing mode, new ways of understanding can be gained: the 'being mode is characterized by a sense of freedom, freshness, and unfolding of experience in new ways' (Segal, Williams & Teasdale, 2002: 74). A theory behind the practice of The Yard was that creating a relaxing, facilitating environment through horses and nature, together with the therapeutic orientation adopted, would lead to relaxation and openness to learning occurring naturally.

Shaver *et al.*, in their paper on the links between attachment theory and mindfulness, talk of the central tenet of 'a stronger and wiser other who helps a client or seeker of emotional stability become less anxious, less avoidant, more secure, and more effectively mindful...' (2007: 269). Through extracts in the following chapters it is shown that the horse may act as the 'stronger and wiser other' and provide some of these characteristics. Another area within the mindfulness literature which has links with EAT/L is referred to as 'authentic functioning'. Heppner and Kernis (2007) talk of how mindfulness leads to 'authentic functioning' and Brown, Ryan and Creswell state that 'mindfully informed action appears less likely to be regulated by ego-concerns, and thus is more likely to represent integrated, *authentic* functioning' (2007: 218). It is suggested that authentic functioning is where the individual operates true to one's core self, with an awareness and self-knowledge, and with 'relational orientation' towards others that includes openness and connectivity (Heppner & Kernis, 2007). The same authors state that their research 'has shown that mindfulness and authenticity are interrelated' (Heppner & Kernis, 2007: 249) and that high authenticity is linked with many of the aspects reported as being important to healthy mental functioning, such as self-strength, coping ability, low stress rates and positive relationships. They go on to claim that becoming more authentic and mindful can lead to lower rates of aggressive behaviour due to the individual possessing greater 'secure' self-esteem, as opposed to 'fragile' self-esteem (Heppner & Kernis, 2007; Heppner *et al.*, 2008). A reduction in aggressive behaviour in adolescents with a history of conduct disorder participating in a mindfulness programme was also reported by Zylowska *et al.* (2008).

The emphasis that mindfulness places on the link between the mind and body has many parallels with EAT/L. In their mindfulness study on adults and adolescents with ADHD, Zylowska *et al.* describe how the training or technique is experiential in practice and that the participants 'learn to reduce arousal through breathing and relaxation exercises and to bring an openness and acceptance to their emotional experiences' (2008: 739). Heppner *et al.* suggest that 'aggressive behaviour is one means by which people attempt to restore their damaged self images' and that 'social rejection or ostracism often leads to heightened aggressive behaviour' (Heppner *et al.*, 2008: 487). An example is given in Chapter 6 of how Cinderella's defence mechanism and aggressive behaviour was triggered during a TH session with the horse Duchess. It may be that the comment by Segal, Williams and Teasdale is appropriate in this instance when they suggest that 'The MBSR approach allows participants to see how negative thoughts and feelings are often expressed through the body' (2002: 60). In this TH session there may also have been aspects of the participant experiencing the *experiential freedom* that Shaver *et al.* (2007) refer to. This is where a person is free to choose their course of action and take responsibility for their choices, together with *organismic trusting* – making decisions based on what feels right (Shaver *et al.*, 2007)

The research site and research methods

The research site, 'The Yard' TH centre, employed many of the techniques, philosophies and working practices aligned to the fields of creative, humanistic and gestalt therapies, such as play and art therapies that have their roots in psychotherapeutic and psychoanalytic traditions. It was important therefore to employ research methods that embraced and incorporated these principles. Being a small-scale research project, a case-study approach was chosen aligned with some of the methodologies and methods of reflexive qualitative research (Denzin & Lincoln, 2002; Finlay & Gough, 2003; Fuller & Petch, 1995) and in particular a psychosocial approach (Clarke, 2002; Clarke & Hoggett, 2009; Hollway & Jefferson, 2000), which is 'guided in some sense by psychoanalytic methods' (Walkerdine, 2008: 343). Being a practitioner-researcher on the project, the site lent itself to the socioanthropological tradition of employing ethnographic, participant observational, research methods to achieve 'thick descriptions' (Denzin, 2001; Geertz, 1973; Laird, 1994).

A psychosocial research position incorporates elements of this approach, having been described as 'more an attitude, or position towards the subject(s) than just another methodology' (Clarke & Hoggett, 2009: 2). Broadly speaking, psychosocial research is interested in both the conscious and unconscious elements that are present in the researched and researcher during the research experience. It employs methods aligned to various psychotherapeutic approaches, such as loose, open-ended questions, reframing or rephrasing questions using the participants own frames of reference, and looking at issues of projection (Clarke, 2002; Clarke & Hoggett, 2009; Hollway, 2009).

Being a practitioner on the study site, I was in the fortunate position of having an in-depth understanding of EAT/L and TH practices, and knowledge of the site and participants. It is suggested that 'you can only adequately describe work in a particular occupational setting if you are a competent member' (Travers, 2001: 80). However, the opposite criticism is also levied – of the problems inherent in 'making strange' a social situation in which the researcher is closely familiar. By being so close to a subject or site there is the danger that knowledge becomes 'deadened' and understanding is taken for granted (Atkinson, Coffey & Delamont, 2003). In line with the epistemological values and philosophical approach and practice of The Yard, a reflexive, heuristic research model located within the psychosocial approach outlined above was chosen (Clarke & Hoggett, 2009; Finlay & Gough, 2003; Moustakas, 1990). It is suggested that heuristic research is naturally aligned to a client-centred therapeutic approach as both are concerned with 'investigation into the nature and meaning of human experience' within a non-directive and, primarily optimistic, Rogerian view of human nature (O'Hara, 1986: 174). These approaches lend themselves to seeking ways of exploring, experiencing and describing the multilayered facets of human experience, with the researcher acknowledged as being an active participant in the research process (Fuller & Petch, 1995; Hertz, 1997; Moustakas, 1990; Riessman, 1994). Within these wide cross-disciplinary qualitative research positions, an ethnographic, case-study design was chosen that employed participant observation with detailed fieldnotes and a combination of semistructured and, latterly, less structured, open, field interviews (Henn, Weinstein & Foard, 2006; Hollway & Jefferson, 2000; May, 2001).

A criticism sometimes levied at the small-scale, in-depth, qualitative research study is that it lacks the ability to provide any meaningful conclusions and apply generalisations of findings. In the field of social work an emphasis on 'evidence-based practice' relying on objective, empirical, medically based research design has been favoured, although not without criticism (Webb, 2001). However, Payne and Williams (2005) argue that what they call 'moderatum generalizations' are possible and, indeed, inherent in qualitative research. The authors suggest that this can be achieved in part by greater reflection on the research process and 'expressing their more modest claims in clearer terms' (Payne & Williams, 2005: 311).

Because the population studied was small, with only seven young people, a case study was the most suitable design as 'the case study, is in many ways, ideally suited to the needs and resources of the small-scale

The Yard: The research setting and participants

researcher' (Blaxter, Hughes & Tight, 1996: 66). In addition, as the research study was interested in providing insight into TH, an instrumental case study was chosen with The Yard as the case, rather than the individual participants, as in psychotherapy research.

The Yard: The research setting and participants

The research setting

The site for the research study was a TH programme located in the UK that was established whilst I was working as a social worker for a private foster care company. Foster care companies recruit, assess and train foster carers with whom they then place young people who have been taken into care. These young people are referred to them from local authority (LA) social services departments. The majority of these young people have been removed from their families and taken into LA care due to physical/emotional/sexual abuse and/or neglect. As a result, many often display emotional and/or behavioural problems and/or mental health issues and are labelled with conditions such as ADHD. In 2009 there were 60,900 young people in care in the UK (DCSF, 2009) and they suffer disadvantage in many areas, not least in educational achievement (Jackson & McParlin, 2006; Jackson & Simon, 2005).

The LA has a duty to assess the needs of these young people and to provide them with appropriate services. In many cases this can include therapeutic interventions, such as counselling, psychotherapy and/or alternative educational provision. However, many find it difficult to engage with traditional methods of statutory mental health and educational provision and so alternative ways of reaching and engaging these hard to reach, excluded young people are sometimes used. Outward bound adventure courses located in the natural rural environment, providing activities such as rock climbing, water sports, rope courses and other outdoor approaches, including 'school green gyms' and 'forest schools', are employed (BTCV, 2009; Moote & Wodarski, 1997; Ungar, Dumond & Mcdonald, 2005). Another emerging therapeutic approach is the expanding field of EAT/L, which incorporates EAP, EAA and the area of this research project, TH. Whilst there are both subtle and larger differences in the various approaches, the common feature is that trained professionals employ horses to provide a therapeutic experience in order to bring about positive benefits for people experiencing difficulty of some kind.

Many of the young people who were referred to the foster care agency were assessed and in receipt of some form of additional intervention, but, in many cases, they found it difficult to engage with these traditional services. With a background, interest and training in the field of EAT/L and, in addition, my own horses, when the suggestion was made to incorporate TH into their range of service provision, it was a relatively easy, albeit very pioneering, transition for the company to embrace the concept and agree to establishing a pilot TH programme to offer to some of these young people. Following guidelines set down by the two main emerging organisations seeking to regulate and professionalise the field of EAT/L in the USA, EAGALA and the Equine Facilitated Mental Health Association (EFMHA), it was policy for two members to facilitate a session. This consisted of a lead practitioner with a professional care qualification (social worker, teacher, counsellor, therapist) and a horse specialist with relevant experience and/or qualifications. In addition, EAGALA and EFMHA stress the importance of both practitioners having dual experience and training in both horse management and child and social care issues in order to provide a safe, therapeutic and meaningful environment. The Yard followed this procedure, organising appropriate training as required.

The location and setting of The Yard

The TH programme was located in a peaceful rural location and consisted of a large open barn for the horses, a feed storage barn and a small office/educational area. Nine horses were involved over the course of the study with between four and six in the herd at any one time. The area employed for the TH programme – 'The Yard' – was leased from the owner of the premises and was a self-contained area positioned in a private area on a larger equestrian facility. The whole farm sat on the side of a peaceful river valley leading into a small market town approximately three miles away, and it comprised approximately 40 acres surrounded by quiet country lanes, tracks and bridleways that are ideal for riding and walking.

The yard for the TH programme was an area of compacted hardcore standing containing a wooden barn of approximately 14 ft × 26 ft. It was an open barn with a gate, which could be left open so that the horses were free to wander in and out at will, or shut in order to contain one or two of them – for example, at feeding time or if one was ill or injured. For training and activity purposes there was a round pen and sand riding arena of 20 m × 40 m situated a short distance from The Yard located next to the main barn and stables of the larger farm. Surrounding and sheltering the yard on its north-eastern end was a small wood consisting of oak and hazel trees in which a large number of birds lived. Jay, woodpecker, a very inquisitive and territorial robin who had made The Yard his home and many other small garden birds were constant companions. In the spring and summer months, cuckoo and swallows joined the menagerie and a buzzard was often heard with its eerie call before it was seen, gliding high above the fields then swooping like a stone to pluck an unlucky fieldmouse or, on occasion, a young rabbit in its talons. In a few TH sessions a young person was privileged with the opportunity of seeing a resident fox trotting silently across the fields, stopping from time to time to check on its surroundings. On those rare and magical moments we sat down quietly on the hill whilst the young person watched intently, time standing still for a moment, until the fox disappeared from sight into the undergrowth. Very occasionally wild deer were seen, having left the safety of the large woods across the valley in order to search for food. These were rare and savoured moments and were remembered for a long time afterwards. One young person asked me several months after we had seen a lone deer standing perfectly still, watching us from the safety of its camouflage against the hedge: 'Do you remember seeing the deer?'

The bench

Back in the yard a rustic wooden picnic table and bench had been placed next to the shed-cum-office where a lot of the TH sessional activity took place. From here was a wonderful view across the valley and down to the small river at the valley bottom and the woods on the hills beyond. On a clear day the estuary river mouth to the south could be seen. Many TH sessions were spent merely sitting and relaxing at the bench, which had been chewed on by the horses and so had seen better days. Drinking tea, eating lunch, and watching and commenting on the horses and wildlife was a popular pastime with many of the young people, who would often seem to relax once they realised that there was no pressure on them to do any particular tasks. As the type of horsemanship practised on the TH programme was aligned to 'natural horsemanship', the horses were free to range in and out of the barn, yard and fields and so often wandered over to us at the bench out of natural curiosity or in search of a snack or neck scratch. Having this set-up served a number of purposes: it was both a philosophical and practical arrangement.

Natural horsemanship

During recent years a 'new' method and philosophy for horse training and management has developed (Birke, 2007, 2008; Latimer & Birke, 2009; Rashid, 2004; Roberts, 2000; Widdicombe, 2008), although the method, whilst 'new' in the sense of it becoming more mainstream and having a huge marketing base in the past ten years or so, is in fact based on principles practised by a number of horse trainers for centuries (Dorrance & Desmond, 2001; Rees, 1984). Natural horsemanship is based upon principles of partnership, kindness and a more 'natural' approach to keeping and training horses, as opposed to the traditional and often violent methods of domination through fear. Instead of being housed individually in separate stables, which is alien to horses, which have very strong social bonds, horses kept under a natural horsemanship approach live in groups in fields where they can create a more similar herd structure to that of the wild horse. To offer protection from the weather, free-range-style barns or field shelters are sometimes supplied which the horses can use at liberty. This was the set-up employed at The Yard in the research project.

Employing a natural horsemanship approach has many advantages over traditional methods on a number of levels. From a practical point of view the time involved in daily horse-management chores can be reduced in many areas, leaving more time for interacting with the horse. The time-consuming task of mucking out individual stables is reduced if the horses are housed in barns, which can be deep-bedded and then cleared out with a tractor at less frequent intervals, in the same way as overwintered cows are housed. At The Yard the horses' barn merely had an earth floor and no bedding, which provides an easier, and naturally draining, surface for collecting horse manure than a concrete base. If a horse needed to be contained in the barn due to illness, for example, a bed of wood shaving was laid. Another benefit of keeping horses free range is a reduction in the time involved in 'turning out' and 'bringing in' horses every morning and evening, in addition to the other chores involved, such as changing rugs over from outdoor to indoor rugs each time.

However, the greatest benefits that many horse owners experience are related to the horses' general well-being, and their mental and physical health. Traditional methods often mean that horses are kept shut in stables for long periods of time, often resulting in injuries, ill health and behavioural problems as a consequence. The system is also cheaper to run.

A good fit: Natural horsemanship, philosophy and practice of The Yard

It was considered that natural horsemanship was a good fit with the TH practiced at The Yard for a number of philosophical as well as practical reasons: it was found that the approach offered a number of advantages for providing a therapeutic and educational environment. As mentioned previously, the horses were free to wander around and display natural behaviours, and because they lived in established herds they had formed bonds and established their 'pecking' order, although this is a contested term (Rees, 2009, 2013). This meant that the dangerous behaviours that can be displayed when horses are constantly moved around or introduced to each other, such as kicking, biting and more general unpredictable behaviours, were avoided. Additionally, due to their living 'free range', they were more relaxed as they did not have the repressed energy which is a consequence of being kept locked in a small space for hours at a time. A natural horsemanship approach understands that being stabled is wholly unnatural for an animal that has evolved to be constantly moving and eating, and can result in unhealthy and dangerous behaviours towards themselves, each other and humans.

Because of all these factors, The Yard was therefore found to be a safe environment in which to introduce young people into the horses' space and allow them the opportunity to observe and interact with horses at close quarters in their 'natural' environment. This seemed to provide a number of benefits, experiences and learning opportunities. The majority of young people participating in TH had experienced abuse, neglect and other deprivation or negative experiences of some sort, resulting in an understandable difficulty in trusting adults. In order to survive and protect themselves they had therefore had to develop survival and coping strategies (Gammage, 2008). This could manifest in emotional/behavioural problems, which could be difficult for themselves and others to manage. It was hoped that by utilising a natural horsemanship approach at The Yard, together with humanistic therapeutic principles of 'unconditional, positive regard' (Rogers, 1951, 1992) and aligned to gestalt approaches, which take a holistic approach to the mind–body link and incorporate body exercises in therapy (Clarkson, 1999; Mackewn, 1997; Perls, Hefferline & Goodman, 1972), that a positive learning and therapeutic environment could be created on a number of levels. First, The Yard strived to model an environment of trust, partnership and respect towards horses as sentient, emotional beings, which it hoped would provide a positive model for the young people.

Second, because horses are large, powerful and command respect, but are also curious and playful, and the young people are in *their* space, the young people were immediately engaged and motivated, and also in a different and, sometimes, perhaps slightly more insecure position than they were used to in their normal environments. The combination of these factors served to engage the young people with the TH practitioners in a way that they may not have in a different, more conventional environment. In addition, there were found to be similarities in the relationship between a horse and a young person to the qualities that are suggested to be important in the therapeutic relationship between the client and the therapist, which were partly attributed to the style of horse management and approach adopted at The Yard (Chardonnens, 2009; Rogers, 1951; Yorke Adams & Coady, 2008). This is explored in more depth in Chapter 5.

The participants: The young people

It was initially hoped that eight young people would participate in the research. Unfortunately it proved impossible to gain signed consent from one of the participants due to his sudden change of foster placements to another part of the country. Finally, therefore, seven young people who attended The Yard participated in the study. Pseudonyms were employed for the sake of confidentiality. In terms of gender differences there were five girls and two boys. This was in contrast with the general trend of the gender divide of participants who attended The Yard – normally more boys than girls were referred. This might have been due to the fact that initially The Yard received referrals solely from the foster care agency and more boys are in the care system than girls (57% boys and 43% girls during 2008–2009; DCSF, 2009). However, at the time of the study The Yard had started to take private referrals and referrals from a number of other agencies, so this may have changed the gender differences.

The longest participants, Minimax and Lucy, came intermittently over the whole two-year period of study. The shortest, Kelly, came for just under six months before she discovered, during a TH session through a phone call from her social worker, that her residential placement was to be brought to an abrupt end, so she was moved to a new foster placement in another area. This provides an example of the disjointed and chaotic nature of many of these young people's lives in care, and of some of the difficulties involved in maintaining a meaningful therapeutic experience between the participants, horses and practitioners. Generally the young people came for a number of months, generally fortnightly or when their timetables allowed. Their ages ranged between 11 and 16 years old, apart from Lucy, who had her 21st birthday during the time of the research study. The participants were understood to be in the 'at-risk' category due to their various psychosocial circumstances. Their referral forms stated various difficulties, ranging from emotional-behavioural and anger-management issues, to low self-esteem and difficulties in forming relationships. In line with the child-centred ethos of The Yard, I note my discomfort with the range of labels with which many of these young people found themselves and of their associated connotations (Jones, 2003).

The young people were generally referred from a range of sources, including statutory agencies, such as social services, youth offending teams (YOTs), a pupil-referral unit and a specialist children's home, although two were introduced to The Yard by individual parents. Two of the young people were referred from the foster care agency which initially funded The Yard. A further two were referred from a therapeutic residential facility. Lucy had been attending The Yard on an informal basis for a number of years due to her love of horses. Her mother said that she had suffered bullying and exclusion at her previous school, and Lucy explained that she had a diagnosis of non-verbal learning disability and was assessed as being on the autistic spectrum. She went on to discuss her physical eye condition, which was part of the reason for her being able to 'see in the dark better than in daylight' and to generally be able to relate better to animals than people. Lucy continues to attend, helping with the horses on a voluntary basis at the time of writing. Emma was referred jointly by her social worker and adoptive mother, and also continued to attend privately on a voluntary basis after The Yard lost its funding. A number of the participants had a Statement of special educational needs (SEN) due to difficulties at school, and one saw an educational psychologist. Wayne was referred by a local YOT as part of their mentoring project. He attended together with his mentor, Peter, who also took part in an interview for the research. Another young person who was to be included in the study unfortunately left The Yard before he had signed a consent form. This was disappointing and frustrating, because consent had been obtained from both his foster carer and his social worker, and he had given his verbal consent. Unfortunately, despite persistent and prolonged attempts to contact him afterwards, this proved impossible due to his having moved placements, together with a change of foster carer, which was a common feature in the disjointed lives of these excluded young people.

Whilst it would have been ideal within a participatory framework for the participants to have all given their own self-descriptions, the reality was that this was only achieved with two of the young people – Emma and Lucy – so it was not deemed appropriate to include only their accounts. Reasons for not managing to achieve self-descriptions from the other five participants range from time pressures in sessions and reluctance of the young people to partake in this exercise, and was also due to the disjointed and transitory nature of their lives, in that they moved on with little or no notice to The Yard. Gaining the young person's trust and finding the appropriate time to approach them and their gatekeepers to discuss the research was a time-consuming and delicate process, which was fraught with difficulty and ethical dilemmas at times. Some of the young people were not entirely comfortable talking about themselves so this was not pursued, leaving it up to them so say if they changed their minds. However, at the other end of the spectrum, Lucy was very articulate and willing to provide a detailed self-description.

The participants: The adults

In order to give a further perspective and depth to the study, a number of the adults involved in The Yard were invited to contribute to the research. This was through formal, semistructured interviews or questionnaires, or both, in addition to informal ethnographic 'field' interviews. The two therapists, Sally and Deana, who practised at The Yard gave in-depth interviews. Additionally, Peter, who was Wayne's mentor from the YOT, participated in an interview, as did Minimax's foster carer, Angie, and Emma's adoptive mother, Linda. Further contributions in the form of questionnaires came from other foster carers, and the manager from Freya and Kelly's residential home.

The horses

Nine horses in total were resident at The Yard during the course of the research study, although some left and others joined during this time. There were between four and six horses at any one time in the herd. They ranged from an elderly Lusitano mare, Duchess, who was very sadly put down due to old age and deteriorating arthritis during the course of the fieldwork, to two young, semiwild horses. In the middle were the constant herd members, consisting of Ruby, a cob × mare, and Jason, an Iberian × gelding (son of Duchess). Their temperaments

TH sessions

Depending on the individual needs of the young person, TH sessions could offer either a straightforward horsemanship approach or a more therapeutic focus facilitated by The Yard's play therapist or counsellor. Alternatively, sessions could incorporate an educational emphasis, such as including quizzes or small projects based on equine subjects. Whilst both the time and length of sessions could be flexible (although more often than not they were dictated by finances), they were generally set at two hours. The normal procedure was for a one-hour assessment session to be followed by a block of six sessions, which were then reviewed. They could then be ongoing or another block of six purchased. In reality it was often the case that the child's circumstances dictated the length of time that they attended, with one young person coming for half-day sessions weekly for six months, another three attending on and off for over a year, and Lucy and Minimax coming intermittently over the whole two-year study period. In some cases young people were referred to attend for the initial six sessions but only attended once or twice due to their foster placements breaking down and their being moved out of the area.

Children and young people were transported to The Yard by taxi, by their carers or, occasionally, by their social worker, who then returned to collect them at the end of the session. Generally only one child participated in a session at a time. This was due to both practical reasons and the practice orientation of The Yard. The children and young people participating were usually in sole foster placements and often referred due to experiencing difficulties and/or disruption with their education. Practically, trying to organise a number of children who may be excluded from school at different times, had different school patterns and lived in different geographical locations would have been extremely time-consuming and impractical. Many of the young people referred had difficulties in their relationships with peers and one of the goals of the TH sessions could be to initially work 1–1 with them, with the aim of working towards group sessions at a later time. This was the case with Minimax and the aim was achieved. The exception to this general pattern of sole sessions was the holiday project. This was where foster carers could purchase TH sessions as a holiday activity at a reduced fee, subsidised by the foster care company. Because the inhibiting factor of school timetables was absent, small groups of two or three young people could more easily be accommodated on the holiday projects. Two of the young people, Minimax and Cinderella, who attended the holiday project participated in the research, although both were also referred through the formal referral process.

Activities in a TH session ranged from initially spending time with the horses, and observing and discussing horse behaviour in the herd, building up to learning how to work with and handle horses on the ground. During the research study, insurance cover was extended to include riding. Therefore if the young person wished to they could participate in an 'invisible riding' session. This is where a young person rode the horse bareback (no saddle) in the round pen (a contained training area) and practised riding with no saddle or reins using the power of intention and body language alone. It was seen to be a very empowering experience for a young person to learn that they could connect with a horse in this very subtle way. Other, ground-based exercises practised were 'join-up', where the young person worked with a horse loose in the round pen, learning how their body language influenced the horse. Through this exercise the young person learnt how to build up a relationship with the horse in order for them to follow them around with no lead rope. For some of the young people who had concentration difficulties and were labelled with conditions such as ADHD, Asperger's/autistic spectrum and behavioural difficulties, horse-based fun activities such as treasure and scavenger hunts and the popular obstacle course were introduced into sessions. These could be completed either leading or riding a horse, depending on ability, confidence and the length of time during which the young person had been attending. Art activities based around the horse were also popular, either painting old horseshoes which the young people took away with them as a memento, or drawing or painting their favourite horse. An American Indian tradition learned on an EAT/L training course of decorating the horses with hopes and dreams written onto small notes then plaited into the horses manes and tails was another activity. Sometimes art sessions would take off in their own direction, and art and play therapy techniques could then be incorporated if this was appropriate and practised by a qualified therapist as part of the experiential approach to sessions.

The consent process, some ethical issues and power imbalances

Gaining consent from the young people to participate in the research was a long and convoluted process. In many cases, three or four

74 *Equine-Assisted Therapy and Learning with At-Risk Young People*

gatekeepers were involved prior to even seeking to discuss the research with the young people. This could include the foster carer, the foster carer's social worker, the child's social worker and then the child's parent if they were on a voluntary care order. Even this process was ethically problematic. How 'participative' was the research if the young people's gatekeepers had previously been contacted before speaking to them personally? There were additional concerns about issues of 'informed consent' which applied to these young people as they were classed as vulnerable, both by being under the age of 16 and because of their status of being in care or considered 'at risk' in some way. For this research project, ethical considerations were paramount and were part of an ongoing process throughout the study. Initially consent was obtained from the university ethics committee, and a child protection policy was placed alongside the other policies and procedures of The Yard. Separate information sheets and consent forms were designed for the young people and adults, and these were given out in advance in order that the young person could read them together with their carer, parent or social worker in their own time. It was made clear that the young person was free to withdraw from the research at any time with no bearing on their attendance at The Yard (see appendices for examples of consent and information forms, and child protection policy).

To attempt to make the process as robust as possible, where feasible the young person was spoken to both individually and together with their foster carer or other appropriate adult. This was in order to give the maximum opportunity for the young person to feel able to voice their opinion as to whether they wanted to participate and fully understood the process. The social worker of one of the young people who attended would not give permission to speak to him about the research. She stated that he had enough bureaucracy in his life already and argued 'What would he get out of it anyway?'

Pseudonyms

For the sake of confidentiality and in line with a participative approach, all of the young people were given the opportunity to choose their own pseudonyms. However, only four of the participants did so as the others unfortunately stopped attending The Yard before having decided on their pseudonyms. The young people's opinions and collaboration were sought on research design, such as questionnaires and interview techniques. This was successful with a number of the participants, such as Lucy, who was very articulate and keen to engage with all aspects of the research, but was felt to be intrusive and inappropriate with others, such as Minimax and Wayne, who seemed to find being asked questions tiresome, so this was respected.

Data-collection procedures and instruments

A 'practice near', practitioner-researcher, model

The method utilised for this research project was as practitioner-researcher within an ethnographic approach. A practitioner-researcher approach has many parallels with participant observation, which, in turn, has its origins in social anthropology. Participant observation requires the researcher joining in the activity or lives of the research participants, getting to know them and their world, and on their terms. This can mean having to learn new social norms, languages and signs in order to become accepted in this world (Corrigan, 1979; Okely, 1983; Foard, 2006: 172). As a practitioner at The Yard, I had gained the 'insider status' described by Henn, Weinstein and Foard (2006). Throughout the research project, many questions were raised concerning the practitioner-researcher role and ethics of 'practice-near' research, where the young people may have easily forgotten my role of researcher in the course of being a practitioner in a TH session. To attempt to counteract this, I would frequently refer to the research and phrase questions to include reference to it, such as 'I wonder if you have anything you could add to xxxx that you think may be useful or relevant to my research'. However, many insights and conversations around the horses would evolve naturally and spontaneously in the course of a TH session, leaving me without the opportunity to raise my research, and, indeed, this would have lost the spontaneity altogether. My position therefore would swing between practitioner and researcher at different times, and was, of course, always subject to issues of power imbalance with the young people due to my position as adult, as practitioner and, in addition, as researcher. Compounding these problems were the further dimensions of issues of power and control inherent in the field of psychotherapy to which The Yard was aligned. Whilst The Yard sought to practice a non-directive, child-centred model which attempted to overcome some of these issues by allowing the child to dictate the direction and pace of the content of TH sessions (within certain safety parameters), it is argued that 'counselling and psychotherapy exist at the intersection of liberation and social control' (McLeod, 2001: 18). McLeod suggests that this is on two levels: first, due to the intensity of the client–therapist relationship, there is the potential for

76 *Equine-Assisted Therapy and Learning with At-Risk Young People*

manipulation and control; and, second, external state and commercial forces put pressure on the therapy field to become agents of social control, a charge also levied at social work (Day, 1981). In the case of The Yard, it could be argued that by practising exercises with horses which could enable young people to become more aware of the effects of their own behaviour, this was behaviour modification on some level, with the connotations of social control and manipulation this perhaps raises.

Whilst there are no easy resolutions to these tensions, a counter-argument to criticisms levied at practitioner-researcher methods is that 'positioning research close to practice creates opportunities for greater complexity of experience to be understood by practitioners, across a wide number of fields' (Hingley-Jones, 2009: 413). Hingley-Jones (2009) goes on to argue that learning about the inner worlds and emotional experiences of service users can lead to a greater understanding of the possible structural disadvantages facing them in their lives. The hope would be that this understanding would then translate into positive action.

Fieldnotes

Detailed fieldnotes written by hand or typed as soon as practically possible after TH sessions comprised the largest and possibly most valuable data-collection method in my research, amounting to hundreds of thousands of words. Atkinson, Coffey and Delamont (2003) suggest that fieldnotes and journals 'are the building blocks of qualitative research, a place for the accumulation of data and reflections' (2003: 60). My fieldnotes evolved from almost clinical, rather objective, short notes at the beginning to long reflexive accounts of the research process based on detailed 'thick descriptions' of the TH sessions that had taken place. TH sessions involved many layers of communication, sensory exploration and different experiences, not least because horses were the major component and communicate in a different way to humans. Rossman and Rallis write that you 'need to turn what you see and hear (or, perhaps, smell and taste) into data' (1998: 137). The experiential nature of TH sessions involved times when the participants would hug and hold the horses, smelling deeply into their necks. How to describe the earthy, musky, yet sweet and individual smell of a horse from a participants' perspective proved to be a challenge. One young person, Minimax, suggested that the mare, Duchess, 'smells like horse perfume'. Therapist Deana described the experiential nature of being with the horses as

'You're just with the horse. Smelling, feeling, riding', illustrating the multifaceted layers involved in the experience for her. I attempted to develop what Hollway (2009) calls 'experience-near' fieldnotes, which seek to 'document the emotional dynamics of research encounters' in order to convey the 'alive quality of the event even long after it had happened' (Hollway, 2009: 465).

Before having the opportunity to write up my own personal fieldnotes and reflexive journal after a TH session, we had, as the first priority, to write up the 'sessional form'. This was completed by both practitioners and was a record of the session. It was then sent on to the young person's social worker or other relevant referrer. In line with a participatory child-centred approach, we would attempt to include and involve the young person in this process, asking what they would like recorded, what had been important for them in that session, what they felt they had learnt and so on. Sometimes this was successful but at other times the young person would not want to engage in this process and often we would merely run out of time, especially if the young person was engrossed in an activity with a horse and it felt inappropriate to cut this short. As these sessional notes formed an important part of the session and were co-produced, they were also an important part of the data collected.

Interviews

Alongside observation and fieldnotes, interviews of various forms were conducted with many of the participants. Blaxter, Hughes and Tight (1996) define the interview as a method involving questioning and discussing issues with people: 'it can be a useful technique for collecting data which would be unlikely to be accessible using techniques such as observation or questionnaires' (1996: 153). Interviewing within the social sciences can take a number of different forms. Denscombe (2003) lists these as structured, semistructured, unstructured, and focus groups and group interviews.

For both epistemological and methodological reasons, for this research project what have been termed 'informal ethnographic interviews' (Spradley, 1979) and 'conversational interviews' Patton (1990) were adopted. This is where the researcher asks questions in the course of the participant observation (Blaxter, Hughes & Tight, 1996; Silverman, 2005). In addition a number of more structured interviews were conducted, although utilising a semistructured questioning style in line with the value base of the study, which sought to explore and allow the voices, feelings and opinions of the participants to be heard, within 'thick descriptions'. These semistructured interviews were conducted with five adults and three of the young people who participated in the research. They were recorded on audiotape (except for one with Cinderella, which was a shorter unplanned interview before a TH session, where advantage was taken of some additional time and so it was handwritten) with permission from the research participants and lasted between half and hour and two hours. Clarke (2002) describes the method of psychosocial interviewing as unstructured in nature and suggest that it is important to avoid using 'why' questions. He suggests instead asking loose, very open-ended questions in order to avoid 'putting words in people's mouths' (2002: 177). A further technique of psychosocial interviewing is closely aligned to principles adopted from psychotherapy and counselling techniques – those of using participants' ordering and phrasing, and rephrasing questions in the 'own words. As noted by Hollway and Jefferson (2000), this takes discipline and practice. An important feature of the psychosocial interview is that it is interested in any projection and unconscious processes that may be present in the interview and of 'Why do people tell certain parts of certain stories? Why are they telling them?' (Clarke, 2002: 177). It was found to be the case that many of the young people would identify with certain horses and would seem to project parts of their characteristics, personalities, behaviours and, sometimes, issues they were perhaps exploring through the horses. A psychosocial approach to interviewing allowed for an exploration and analysis of these processes after the interview, and to return to the participants where possible for clarification.

Questionnaires

The third data-collection technique was a short questionnaire which was given to both the young people and the adults (see appendices for examples of these). Whilst questionnaires are more usually associated with quantitative research methods (D'Cruz & Jones, 2004), it was felt that it was appropriate to incorporate them in this study due, in part, to their being a possible useful medium for the young people to participate in the research. The questionnaires were adapted and evolved from feedback questionnaires already devised for the TH programme as part of the evaluation process of The Yard. The young people were invited to assist in designing these questionnaires and it was found to be one of the areas that some of them felt most able to participate in, Lucy and Cinderella in particular being very imaginative about thinking up possible questions. However, this also had its drawbacks as some of the questions that they suggested were not ones that I considered to be useful or appropriate for the research, so this caused me a dilemma. How could I, on the one hand, ask for their input but then, on the other, only want input which suited me? An example of a question Lucy suggested for the questionnaire was 'How old is Duchess (a horse at The Yard)? This clearly had no relevance to the research but was obviously considered important by Lucy. In the end this was resolved by designing a new quiz for the TH programme which incorporated some of these questions.

The questionnaires were given to the young people and adults to complete in their own time and either handed back at our next meeting or mailed at a later date. This was to avoid directly engaging or intervening with the participants as they completed the questionnaire (May, 2001). The questionnaires were as brief as possible as it was acknowledged that the majority of the young people who participated in TH disliked writing and completing forms. In one case a foster carer wrote the answers to the questions on the young person's behalf as he said he didn't want to do the writing but was happy to share his experiences. Questions which were as open-ended as possible were deliberately employed, such as 'How does being around the horses make you feel?' This was in order to encourage responses which were, in return, as open-ended as possible, although it is recognised that the question does imply that the horses did make the participants feel something. Interestingly, this question elicited a similar reply from many of the participants, both child and adult, which was to incorporate the word 'calm' in the answer in some way.

Data analysis procedures

Formal initial analysis and transformation of the data began by transcribing the interviews. A thematic open coding analysis procedure was employed where the data were gathered and organised by initially looking for similar words, themes and patterns (Braun & Clarke, 2006; Coffey & Atkinson, 1996). This process eventually uncovered a number of categories which were then organised into two main thematic blocks of social well-being and psychological processes, although there was clearly overlap between the two. In line with a participative approach, the results were shared with a number of the participants in order to verify my interpretation of their experiences, and to seek further clarification on certain questions that had been raised for me whilst sifting

80 *Equine-Assisted Therapy and Learning with At-Risk Young People*

through the data. Unfortunately this was only possible with four of the young people due to the nature of the TH programme, which, due to political and financial reasons, lost its funding during the final year of the study. Also, more often than not, the young people moved foster placements with little prior notice, as shown in Chapter 3 on Kelly's last day. However, drafts of my initial notes from the interviews were sent to a number of the adult participants for their feedback, which was then incorporated into the analysis.

Some problems of the practitioner-researcher position and power issues involved in taking a reflexive and participatory approach in research with children

A reflexive approach to research and analysis understands that the researcher is *part of* the social reality that is under observation. There is no point in attempting to be a detached observer as the process of the research is interactional. My position as researcher, practitioner and white female, together with my background and influences, will impact upon the research process to create an experience that is unique to those particular circumstances (Hertz, 1997). In this research study, I was a practitioner and therefore extremely close to the subject matter and site. A criticism levied at being a researcher-practitioner could be that it doesn't account for 'the problem of "making strange" the familiarity of social worlds from within one's own culture' (Atkinson, Coffey & Delamont, 2003: 25). The problem that this perspective identifies is that the insider finds it difficult to 'see' what they already 'know' inside out. This is something that did arise in discussion with other non-practitioners in the analysis of my fieldnotes where I had assumed some equine knowledge and issues related to child development and TH practice, for example, as obvious so had not seen the need to elaborate on certain aspects.

There are also real issues of unequal power relations. Above the fundamental inequalities between children, young people and adults, 'our structural framework ensures that children and young adults, are marginalized both socially and politically' (Pugsley, 2002: 21), there was the additional fact that I was a practitioner on the research site, so the children and young people who attended TH therefore saw me as a 'teacher'. Clearly, therefore, in this study I was 'not only perceived by the pupils as being socially more powerful, I *was* more powerful' (Pugsley, 2002: 21). I had access to a 'captive' research group and could choose who to invite to participate and how to 'interpret and represent the data

I collected' (Pugsley, 2002: 21). In order to attempt to address some of these issues and in line with my alliance to a participative, reflexive, critical qualitative inquiry approach (Davies, 1998; Denzin, 2002; Denzin & Lincoln, 2002), I hoped to conduct research that was participative and reciprocal and therefore empowering in principle. However, I am also aware of the limitations and contradictions inherent in the concept of 'empowerment'. Humphries suggests that 'emancipation cannot be conferred on one group by another' (1997: 5), and that researchers, being part of an establishment of hierarchical relations, are 'inevitably implicated in power, so that our efforts to liberate perpetuate the very relations of dominance' (Humphries, 1997: 6).

These are issues that raise additional questions and concerns when working with young, vulnerable people. Whilst this study was concerned with attempting to record the experiences of young people participating at The Yard, it could be argued that by attempting to create a climate of participatory research, encouraging the children and young people to express their experiences and give them 'voice', this may have had the implication of raising their expectations of having their hopes, wishes and views taken into account outside The Yard. This is often contrary to their real-life experience within the 'looked after' system of social work (Holland, 2001). Despite the advances in childcare legislation and guidance set out to improve young people's position in the care system, these have only slowly been put into practice (Holland, 2001), and a sociology of childhood understanding remains theoretical and academic in the most part (Christensen & James, 2000; Farrell, 2005; Greene & Hogan, 2005). There are also the implications of issues of disclosure. As has been discussed previously, a humanistic, client-centred, empathetic environment between participant and researcher-practitioner may hold the possibility of the child or young person feeling 'safe' enough to disclose certain past or perhaps present, traumatic or abusive events. Whilst I was a qualified social worker and had received training and supervision in this area, with a child disclosure policy in place at The Yard, issues of 'blurred boundaries' could arise. Perhaps the child or young person could feel 'safer' and more 'empowered' in an environment where they were being encouraged to participate and share their feelings and experiences about the research project. Could this result in them not seeing me as a practitioner and therefore an 'authority' figure in the same way, but as researcher, so further removed from their world of foster carers, social workers, therapists, YOTs and the like? Could this then have the effect of my being seen as more 'aligned' to them in some way and so therefore 'safer' to talk to and possibly 'disclose' to? What could this mean in terms of implications for the research project?

In reality, many of my fears were unfounded; in fact the opposite was sometimes found to be the case. By taking on the dual role of practitioner-researcher with the added layer of bureaucracy this entailed in terms of explaining the research, signing consent forms and so on, this sometimes seemed to have the opposite effect of distancing the young people. One young person, Wayne, repeatedly gave the one word answer of 'Yarp' in a funny accent whilst I attempted to explain the research and consent process to him, and seemed disinterested and distracted. At one point the horse-practitioner, Donna, working with me that day interrupted to ask Wayne 'is that from the film *Hot Fuzz*?', to which he replied 'Yarp', and they both burst out laughing to my and his YOT mentor, Peter's, confusion, highlighting the gap in our knowledge of popular films. However, this interruption seemed to provide a break in the discussion around the somewhat dry and serious nature of the research consent process, and, following this, Wayne seemed to engage more fully by asking a few questions and telling me he was happy to participate, although adding, as did Minimax, 'as long as I don't have to do any writing!'

3

Development of Self-Confidence and Self-Efficacy, and the Opening Up of 'Positive Experiences' and 'Positive Opportunities' through Therapeutic Horsemanship

Introduction

As previously outlined, the majority of the young people who were referred to The Yard could be understood as being 'at risk' in terms of their various psychosocial disadvantages. Much literature supports the suggestion that exposure to certain risk factors in childhood can result in poor developmental outcomes in later life (Howard, Dryden & Johnson, 1999; Hughes, 2004; Rutter, 1985). Many of the young people participating in the research study were in foster care because of being removed from their parental homes due to neglect and/or abuse issues. However, why some individuals appear to overcome adversity and remain resilient despite experiencing these childhood risks, and what traits or factors were common to these individuals, has also been the focus of much research (Cicchetti & Rogosch, 1997; Masten, Best & Garmezy, 1990; Rutter, 1999, 2006; Werner, 1993). The definition of resilience suggested by Masten, Best and Garmezy is 'the process of, capacity for, or outcome of successful adaptation despite challenging or threatening circumstances' (1990: 426). Many of the themes to emerge from this study had links to the literature on risk and resilience: themes such as self-confidence, self-esteem, self-efficacy and mastery, 'acts of helpfulness' (Howard, Dryden & Johnson, 1999), empathy, and the opening up of 'positive opportunities' (Rutter, 1999).

A large body of literature suggests that exposure to early stressful and traumatic events can result in emotional and behavioural symptoms

84 *Equine-Assisted Therapy and Learning with At-Risk Young People*

in children and young people, such as hyperactivity, mood swings, low self-esteem, depression, and other social and emotional difficulties (Bannister, 1998; Egeland *et al.*, 2002; Gunner *et al.*, 2006; Hughes, 2004; Levine, 1997; Masten, Best & Garmezy, 1990). Other authors propose that physical illnesses and conditions arising from a decreased immune system, such as digestive and skin disorders, can also be a result of childhood stresses (Levine, 2005). It was certainly the case that many of the young people who attended The Yard displayed a variety of these emotional, behavioural and physical conditions. Labels of ADHD, depression, self-harm, violent and aggressive behaviour, and conditions such as eczema and asthma were often noted on their referral forms. In addition, many of the participants would seem to suffer from various aches and pains. Kelly, for example, would often complain of headaches and want to just sit quietly in the shade, and Emma would sometimes not be able to ride due to the eczema on her legs being too inflamed and painful. This chapter looks at how the young people appeared to demonstrate that they experienced gains in self-confidence, self-esteem and self-efficacy through being with the horses. Links to how the horses seemed to provide motivation to learn, to regulate their behaviour and emotions, and, in turn, to enable some of the participants to engage in further educational and other 'positive opportunities' are also explored.

'Negative life outcomes'

The young people who were referred to The Yard from the foster care agency had been taken into care due to concerns for their health and well-being; it was understood that they were 'at risk'. As was explored in more detail previously, it is argued that exposure to certain risk factors can result in 'negative life outcomes' in later life. Some of the EAT/L research, together with the literature concerned with outdoor and adventure activities, proposes that therapeutic interventions, such as EAT/L and TH, can be useful in providing at-risk young people with some of the resilience and protective factors necessary to avoid these possible negative life outcomes (Ewert, 1987; Hayden, 2005; Smith-Osbourne & Selby, 2010; Ungar, Dumond & McDonald, 2005).

To date, research on risk and resilience has tended to concentrate on looking at resilience within certain parameters. These have varied from a wider structural perspective of the effects of socioeconomic status, for example (Bronfenbrenner, 1986; Howard, Dryden & Johnson, 1999; Werner & Smith, 1992), to neurobiological (Gunner & Quevedo, 2007), developmental and behavioural understandings (Cicchetti & Rogosch,

1997; Lazarus, 1993; Masten, Best & Garmezy, 1990; Skinner & Zimmer-Gembeck, 2007; Werner, 1993). Whilst it is understood that all of these bring important contributions to the knowledge base, Ungar (2004) suggests that the most important factor is how young people define their own health, and that resilience and risk factors change over the course of life. By employing a child-centred, flexible, experiential approach, The Yard aimed to provide an environment where young people could gain some of the resources necessary to enable them to lead more successful and positive lives in ways which were meaningful and personal for them as individuals, and so a 'dynamic state' model is the most appropriate within this context (Cicchetti & Rogosch, 1997; Place *et al.*, 2002).

Horses and mastery, self-esteem and self-confidence

Within the risk and resilience literature the development of self-confidence and self-esteem is considered to be of major importance, and necessary for both mental health and in raising self-efficacy in at-risk young people. Self-esteem and self-confidence are neatly summed up by Sempik, Aldridge and Becker who write:

> Self-esteem refers to a general feeling of self-worth or self-value, while self-confidence is the feeling that an individual is likely to succeed in a task and has few hesitations or reservations about attempting it.
>
> (Sempik, Aldridge & Becker, 2005: 90)

Having a sense of power and control over one's actions and future is known as self-efficacy (Bandura, 1982) and is clearly linked to having a sense of self-worth and self-esteem. The participants referred to The Yard who were in care largely had little control or involvement in decision-making about their lives and futures. This, together with their traumatic experiences, often resulted in their referral forms stating that they possessed very little self-esteem, and we certainly witnessed low self-efficacy in terms of their limited beliefs in what they could achieve with the horses. If a culture of helplessness and hopelessness with no sense of future is considered to be one risk factor in successful development and life outcome, then acquiring a sense of mastery and feeling of having some control over one's destiny could perhaps promote resilience (Bandura, 1982; Cicchetti & Rogosch, 1997; Katz, 1997; Masten, Best & Garmezy, 1990). Where individuals experience social exclusion on some level – such as many of the at-risk young people referred to The Yard who were excluded from school, were in between numerous foster placements, and possessed emotional and/or behavioural difficulties which resulted in them having problems in forming positive relationships – the effects are multifaceted. One result is that many of these young people get left behind the developmental and educational milestones of their peers, with the additional stigma and effect on self-esteem and confidence that this causes. When they first arrived at The Yard we found that many of them demonstrated very little confidence around the horses, often being quiet and withdrawn, and finding it difficult to engage and communicate. However, it seemed that as they got to know and, correspondingly, trust and feel safe with the horses, they started to gain confidence. The following examples with Minimax illustrate how he grew in confidence as he got to know the mare Ruby and learn how to care for, manage and ultimately ride her.

> Minimax arrived with his foster carer, Angie, who brought him down into the yard. It was a hot sunny day and the horses were taking advantage of the shade in the barn, standing quietly swishing the flies off each other with their tails with an occasional snort or shuffle of hooves. We took Minimax into the barn to meet them, introducing him to them one by one – Minimax, however, remained rather quiet and withdrawn at first, initially standing behind Angie.
>
> (fieldnotes)
>
> Minimax still rather quiet again today after Angie left him at the yard.
>
> (fieldnotes)

It was found that this remained the pattern with Minimax the first few times he attended. However, once he had become more competent around the horses, which for him appeared to be very much linked to developing a bond with Ruby, this seemed to correspond with him gaining confidence in his communication with the practitioners too. He started to become able to ask questions and, importantly, to ask for help with tasks that he found difficult, something he had struggled with initially.

> Minimax's last session of the summer holidays and he asked if he could go for a hack on the lanes, wanting to ride Ruby again. He has grown enormously in competence and, correspondently, confidence, in the time he has been coming and today managed to halter, groom and pick out Ruby's hooves without assistance for the first time. He

had especially struggled with putting on head collars previously, finding the dexterity and organisation required for this very challenging.

(fieldnotes)

Through becoming more competent and managing to master difficult tasks with the feelings of accomplishment gained from this, it may be that Minimax was feeling a sense of mastery and related growth in self-esteem. The literature on risk and resilience discusses at length the importance of having a sense of control over one's actions and environment, which so many of the young people who attended The Yard lacked. Writing about the benefits of AAT, Kogan *et al.* suggest that

An increased sense of control over the environment, stimulated by activities that provide success and praise for appropriate behaviour, can decrease immature acts and learned helplessness.

(Kogan *et al.*, 1999: 119)

The next section looks at how his gains in confidence, mastery and competence with the horses enabled Minimax to transfer his skills to another young person attending The Yard.

'Acts of helpfulness'

Gaining confidence in his abilities and accomplishments seemed to correspond with 'acts of helpfulness' for Minimax. By developing some feelings of self-esteem through becoming more accomplished with the horses, Minimax began to become more helpful and open towards others. This, in turn, then appeared to give him even more confidence. One of the things both Minimax's foster carer, Angie, and his social worker said that he especially struggled with when he first started coming to The Yard was the school environment, and establishing and maintaining relationships with peers. This was perhaps unsurprising considering the exclusion, inconsistency of relationships, and separation and loss he had experienced having been taken into foster care and separated from his brothers. A vicious circle can be established where a combination of emotional and behavioural problems, together with large gaps out of education due to exclusion and the care process, can result in young people getting behind academically. Behavioural problems can then be exacerbated as the young person feels academically inferior, which impinges on already low self-esteem. Jackson and McParlin report shockingly low academic achievements of children in care, with only 5% of care leavers entering higher education (Jackson & McParlin, 2006; Jackson & Simon, 2005). Angie explained that Minimax would get extremely upset in response to any reference to his mother or brothers, resulting in his getting into fights and being excluded from school on a regular basis. However, after he had been attending The Yard for a number of months, he managed to communicate successfully to another young person at their first session how to look after the horses, clearly being proud of his knowledge and ability. His foster carer commented on how this had contributed greatly to him experiencing a feeling of competence and confidence in his own abilities.

> Minimax was very welcoming towards the new boy and didn't appear arrogant or know-it-all, or jealous and possessive at all today, which he has done previously with other young people. He seemed very generous towards this boy and accommodating, showing him how to do the tasks involved in looking after the horses and telling him a little bit about them all, 'he's a grumps sometimes, you got to get to know him first' pointing at Jason and, introducing the boy to Leo, 'he's old and needs more looking after . . . and sometimes he gets ill from eating too much grass'.
>
> (fieldnotes)

A 'target' for Minimax attending TH was for him to move from a 1–1 situation to participating in sessions with one or two other young people. From this short extract, it can be seen how demonstrating his knowledge to the new boy helped Minimax to achieve this and, in turn, grow in confidence. Furthermore, this experience gave him an opportunity to display an 'act of helpfulness' towards another young person. As Minimax normally found it difficult to engage with his peers, and could be very jealous and possessive of his things, and of people, his foster carer, Angie, explained that this was a huge achievement for him. When she was asked what she felt Minimax had gained most from TH she replied:

> Confidence, confidence, knowledge, confidence that he could do it, boosts his self esteem, 'cos he could do it, I can do that, I can show someone how to do that, I know how to do that, I know how to put this on.
>
> (Angie, foster carer)

This was something that the counsellor, Sally, mentioned when the same question was put to her:

Confidence. I think showing others what he has learnt about horses has made him feel good about himself, made him feel that he can do something and he can do it really well. And the fact that he can demonstrate it to other people, you know, it's a big confidence and self-esteem boost, really. And that makes him more relaxed and more cheerful, so that has, you know, a knock-on effect, really.

(Sally, counsellor)

The resilience literature refers to 'acts of helpfulness' (Katz, 1997) or 'required helpfulness' (Werner, 1993), where helping others can lead to positive benefits for an individual. It is suggested that having an area of responsibility which involves caring for something or someone who is dependent on, and benefits in some way from, that care, including animals, can lead to growth in self-esteem and self-efficacy (Kogan *et al.*, 1999; Myers, 2007; Werner, 1993). Acts of helpfulness can lead to the individual learning that they can effect positive change and have something valuable to offer (Howard, Dryden & Johnson, 1999; Katz, 1997; Werner, 1993). In their randomised trial with people with a chronic disease, Schwartz and Sendor (1999) found that those helping others gained greater psychosocial gains than those receiving the help, concluding that 'helping others helps oneself'. Reported benefits from the helping role included improvements in self-confidence, self-esteem, depression and self-awareness (Schwartz & Sendor, 1999).

It could perhaps be seen in the example involving Minimax that some of these benefits might be gained from the caring and responsibility involved in looking after horses, and these in turn transferred to other areas of psychosocial benefit.

'Get to know her first': Confidence and relationship

With Kelly, it seemed that overcoming her fears was also linked to building up a relationship with a special horse first, and was the catalyst for her gaining confidence. Although she told Deana, her therapist, that she had previously attended a riding school and was an experienced rider, initially Kelly was very withdrawn and uncommunicative as the following sessional notes demonstrate:

Initially Kelly was quiet and rather withdrawn as we introduced her to the yard and horses. It was a warm, sunny day and so we sat at the table and chairs in the yard with the horses loose milling around

90   *Equine-Assisted Therapy and Learning with At-Risk Young People*

and eating from their haynets. After we had sat and watched and discussed the horses for a while I asked Kelly if there was a horse that she was particularly drawn to and wanted to work with and groom. Initially Kelly's response was 'I don't know', which I was to discover was often her default answer to questions when she was unconfident.

(fieldnotes)

Deana was keen for Kelly to ride if she wanted to as Kelly had told her that she wanted to do this in their individual therapy sessions previously to attending The Yard. However, when I asked Kelly if she would like to try 'invisible riding' bareback on Ruby, Kelly answered that she was unsure about this telling us, 'I'm not sure...I haven't ever ridden without a saddle before'. When I assured her that she could have a saddle if she preferred, Kelly still declined and said 'um ...I think I'd like to get to know Ruby a bit better first'.

(fieldnotes)

Kelly obviously felt that it was important for her to 'get to know Ruby' better in order to trust her enough to perhaps try riding her without a saddle. As Kelly attended more sessions and did indeed get to know Ruby more, she started to develop the self-confidence to try new activities with the suggested benefits to self-esteem. It is argued that self-confidence and self-esteem can be developed by succeeding in activities which are motivating and challenging (Berman & Davis-Berman, 1995; Carns, Carns & Holland, 2001; Ellis, Braff & Hutchinson, 2001; Katz, 1997; Ungar, Dumond & Mcdonald, 2005). In a TH session a few months later, Kelly gained the confidence to ride Ruby out on the lanes. In this session the difference between Kelly on the ground and once riding on the horse was quite remarkable, as is described in the sessional notes by her therapist:

Kelly seemed quite quiet when she arrived. She groomed Ruby during which time she spoke very little. It wasn't until Kelly was sitting on Ruby, going out for a walk in the lanes, that she really began to chat. She relaxed, smiling, even laughing at times. Hannah and Kelly chatted about Ruby, what Kelly likes about Ruby in particular. She said she feels she has a special connection with her and that Ruby has an understanding of how she [Kelly] is feeling. It really was quite remarkable the difference between Kelly on the ground, to Kelly on the horse. We [researcher and therapist in debrief] spoke about it together after Kelly had gone, wondering whether it is something to do with being higher, not needing to make eye contact, in addition to the confidence gained from riding.

(Deana, therapist: therapy sessional notes)

In their study with young people in care, Ross *et al.* (2009) talk of how the young people would open up and engage during car journeys and guided walks. They suggest that these 'mobile methods' can sometimes seem to give young people the space to explore sensitive topics at their own pace. In terms of the car, they refer to how the seating arrangements direct the gaze forwards instead of directly front on, as does being on top of a horse, as Deana mentions above (Ross *et al.*, 2009). Sibley has called the car a 'protective capsule' (Sibley, 1995). It was found that many of the young people would seem to open up and become more communicative once on the horse. This appeared to be especially so with both Emma and Lucy, in addition to Kelly. Overcoming her fear of riding Ruby appeared to give Kelly self-confidence which helped her to relax and, in turn, communicate more openly with the adults at The Yard. Being able to communicate and have 'social competence' is argued to be one of the internal characteristics of resilient children according to some authors (Born, Chevalier & Humblet, 1997; Cicchetti & Rogosch, 1997; Howard & Johnson, 2000).

'Queen of the world': Empowerment, identification and opportunity

In addition to themes of opening up, which are explored in more depth in chapters 4 and 5, by engaging in activities with the horses which were considered challenging, the participants recorded feelings of achievement and empowerment. For Minimax this was expressed by his exclaiming after a particularly successful session: 'I've had two times of feeling really strong now!' The participant Lucy described that it was the riding which was an important element for her because it made her higher up. Talking about how riding Duchess made her feel, she explained:

She [Duchess] kind of made me feel like, you know, I'm the queen of the world kind of thing because I was higher up.

(Lucy, participant)

92   *Equine-Assisted Therapy and Learning with At-Risk Young People*

Many of the young people who attended The Yard could be seen to be at risk of suffering increased hazards during adolescence. Having suffered exclusion, separation from family and place, and other negative experiences which could result in them demonstrating difficult and defensive behaviours, there was danger of their being unable to form meaningful friendships and relationships, as demonstrated by Minimax. It may be that participating in TH and the consequent benefits of building up relationships with the horses and others, together with gains in confidence, contributed to their gaining feelings of 'identification' with something. It is suggested that during the difficult time of adolescence a significant 'protective factor' is having 'high identification' with peer groups (Compas, Hinden & Gerhardt, 1995; Erikson, 1995; Hampel & Petermann, 2006).

Jackson, Born and Jacob make the claim that

> High identification leads to greater emotional and informative support from the group, whilst low identification is associated with a more marginal status. The latter group tends to have lower levels of self-esteem and cope less effectively.
>
> (Jackson, Born & Jacob, 1997: 611)

In addition to peer-group identification, some of the literature acknowledges that, in the absence of a secure attachment to a loving parent, the next best thing is for the young person to have a 'scaffolding' or network of other positive supportive relationships within the community (Gilligan, 2004). It may be that the horse contributed to some of these aspects by providing a relationship which the young people perceived as supportive and unconditional, which together with identification with a 'mainstream, leisure' activity (Gilligan, 1999) could help them to gain skills, confidence and greater self-esteem; feel useful and 'helpful'; and ultimately form relationships with others.

Interestingly, when the issue of exclusion during her childhood was discussed in more depth with Lucy, she explained that she didn't feel that she was particularly excluded from activities at school, saying: 'I didn't want to do the same things as the "cool" girls anyway.' This presented me with some difficulty, as Lucy had previously said that she had been bullied and that she had moved to attend a different, more alternative, school where it was hoped that she would be happier. However, Lucy did go on to elaborate that she felt more excluded from things as she got older and was looking for a part-time job and work experience placements. She explained that she felt that 'my [unidentifiable word] voice makes me sound like a little girl' when she rang up for jobs and that this put her at a disadvantage in the job market. It went on to be the case that three of the participants – Lucy, Emma and Wayne – chose college courses and work-experience placements that were animal-based. This could suggest that by 'identifying' with the horses, which perhaps provided some of the protective factors suggested as supplied by acceptance and identification with peer groups (Jackson, Born & Jacob, 1997), they were able to overcome some of the negative aspects of their childhood experiences and engage in what Rutter (1999) terms 'positive opportunities'.

Overcoming fears: Risk, challenge and confidence

Some authors argue that engaging in activities which carry an element of risk is necessary in order to grow and develop self-confidence, self-esteem and self-efficacy (Ball, Gill & Spiegal, 2008; Rosenthal, 1975; Voight, 1988). Voight suggests that 'Confronting fear and overcoming it with the help of a partner or an entire group of people can be used to develop self-confidence, self-esteem, group-cohesiveness and decision-making' (Voight, 1988: 59).

Overcoming a fear or lack of confidence in order to participate in an activity with Hector in the round pen resulted in Freya experiencing a huge change in her whole body language, attitude and way of communicating, as demonstrated in the following extract from fieldnotes:

Freya and Hector: 'he followed me like my shadow'

There has been quite a long gap since Freya last attended TH so I ask her if there is anything she would particularly like to do with the horses today. Freya seems unsure shrugging her shoulders and saying 'don't know', which she often does in response to questions. While we are standing by the shed talking the new horse Hector walks purposely over to us and heads straight to Freya, standing close to her and placing his muzzle in her hands. I comment that Deana and I had been in the yard all morning sitting at the bench having a meeting and Hector hadn't come over once! Freya seems pleased with this and we witness a small smile. We suggest that as Hector seems to want to be with her that she may like taking him up to the round pen to do some leading and perhaps try join-up with him. Again Freya is noncommittal and shrugs her shoulders, saying 'well maybe later ... I'm

94 *Equine-Assisted Therapy and Learning with At-Risk Young People*

not sure'. When Freya first attended The Yard she displayed rather a lack of awareness around the horses and was happy to ride on her first session, to our surprise. I wonder if it is a combination of the bigger size of Hector [he is a large, weight carrying cob] and her having gained a bit more awareness of the potential danger of the horses from almost being trampled on previously, that is making her a little hesitant today. Once we have finished grooming the horses we ask Freya if she would like to lead Hector up the track to the round pen but she replies saying 'I don't think I will lead him up, I will let her [Deana] do it'. We reassure her that is absolutely fine and it is sensible to get to know him better first.

At the round pen Freya is initially hesitant to try anything and refuses to go into the pen. However, she is happy to watch whilst I give a short demonstration of how to achieve 'join-up' with the horse, which is where the horse follows you of their own free will with no lead rope. A horse will only do this if it wants to be with you, you have demonstrated some understanding of body language, and that you are trustworthy and safe. Finally, after watching intently, Freya agrees to join me in the round pen, although refusing to lead Hector without me walking beside her initially. Then, after a while she plucks up the courage to lead him alone and I hand her the lead rope. Hector is just as forgiving and generous with Freya in the round pen as he was in the yard, following her closely both with a loose rein and then without the lead rein, around the cones and over the poles laid out, joined to Freya like glue. It is a joy to watch them together. At the tarpaulin Hector is playful, pawing at the blue plastic sheet with his hooves which makes Freya laugh. Then rain starts to come down again and everyone is getting cold, so I suggest Freya practice seeing if Hector will follow her in trot to warm them both up. Freya's expression becomes momentarily unconfident again and she hesitates, saying 'I'm not sure...' so I quickly suggest that Deana and myself accompany her. Freya visibly relaxes and I tell Freya that if she starts to run Hector might follow her, which he does beautifully. A grin spreads over Freya's face again and she exclaims 'I can't believe I am doing this!'. Freya is happy to lead Hector back down the track this time and seems more confident with him now. When he goes to try and put his head down to eat grass, Freya manages to read his body language and prevent him from eating which is a big achievement with a large, strong horse such as Hector. Later, back at the yard when the staff member from The Elms arrives to collect her,

Freya animatedly tells him about the experience and of how Hector 'followed me like my shadow'.

(fieldnotes and sessional notes)

From the example above with Freya it was clear that her whole attitude and body language changed remarkably from the beginning of the session when she arrived, withdrawn, unconfident and uncommunicative, to after she had succeeded in a slightly challenging and engaging task with the horse, Hector. It appeared that the connection with Hector initially facilitated a change in Freya's mood and this led to her wanting to start to participate in the session. Once she had felt the sense of achievement from overcoming her initial fear and experienced the connection of Hector following her of his own accord, this seemed to give her a huge boost of confidence which then enabled her to start to believe in her own self-efficacy, witnessed when she was able to control Hector's behaviour on the way back down to the yard. This new sense of achievement extended into Freya being more animated and initiating conversation about her experiences with the staff member who came to collect her – a very different Freya than the one who had arrived at The Yard only two hours previously.

'Positive experiences', transferable skills and the opening up of 'positive opportunities'

An aim of EAT/L and TH is to equip young people with skills which encourage a sense of self-esteem, in addition to the self-confidence gained from overcoming fears and challenges by participating in activities with the horses. The hope is that these traits can then be internalised and taken into their everyday lives. This was demonstrated by the participant Lucy, who spoke of how the confidence she gained from being with horses could be applied to other areas of her life. She explained that an understanding of horse behaviour and how she needed to behave around horses resulted in her being able to feel more confident. The following extract is from an interview with her:

Lucy: Um, when I'm angry, they make me feel a lot calmer because you have to be calm around them.
HB: Hmm, yes.
Lucy: Otherwise they pick up on you.
HB: Absolutely.

Lucy: And then you won't get such an enjoyable ride. So if you go in with a cross mind and stamp around, then they're not exactly going to be very helpful to you, are they?

HB: No.

Lucy: If you, if you play the opposite of what you actually feel, with me I start feeling the opposite way.

HB: Ah, I see, so you start acting it, then you start feeling it?

Lucy: Yes. It really helps with confidence and things.

HB: Right. So if you sort of act confident . . .

Lucy: I get confident.

From learning about horse behaviour it would seem that Lucy had learnt how to affect her own behaviour in a similar way to how CBT emphasises how thoughts can affect actions (Beck *et al.*, 1979; Lazarus, 1993). CBT and solution-focused brief therapy work by

> teaching individuals to become more aware of thoughts and feelings and to relate to them in a wider, decentred perspective as 'mental events' rather than as aspects of the self or as necessarily accurate reflections of reality.
>
> (Teasdale *et al.*, 2000: 616)

The literature on risk and resilience notes how developing successful coping strategies and overcoming challenges can strengthen the ability to cope in the face of future difficulties (Cicchetti & Rogosch, 1997; Luther, Cicchetti & Becker, 2000; Place *et al.*, 2002). It appeared that Lucy had gained a sense of self-efficacy from learning how to control her behaviour and actions around the horses, which she could then apply to other areas of her life that she found challenging.

A similar example of how learning about horses helped a participant gain a wider perspective of relating horse behaviour to his own actions is given by Minimax in Chapter 5. In learning about how it was necessary to control Ruby and not allow her to stop and eat grass from the lane verges due to possible danger from cars, Minimax was able to apply this to areas of his own life. We discussed the reasons why it was important for the horses to listen to people and that, although they might not always understand the reasons, we were responsible for their welfare. Minimax seemed to understand this and drew an analogy with when he would dawdle on his way to school and, although he didn't go as far as saying that he thought school was good for him (!), he did tell Ruby 'it's for your own good', whilst preventing her from eating the grass. Whilst it was unclear how much Minimax was consciously able to apply some of these metaphors to himself due to his age, it may be the case that he was unconsciously able to do this in the same way that Jones suggests that unconscious connections are perhaps as, or more, powerful for clients (Jones, 1996). Regardless of whether the processes are conscious or unconsciousness, it may be that learning how to relate experiences learnt by gaining an awareness of horse behaviour and applying these to their own experiences, the young people can gain some of the 'protective factors', such as self-efficacy and a sense of mastery, described in the resilience literature (Cicchetti & Rogosch, 1997; Katz, 1997; Luther, Cicchetti & Becker, 2000; Masten, Best & Garmezy, 1990; Place *et al.*, 2002).

Opening up 'positive opportunities'

A number of authors mention the importance for 'at-risk' young people to have exposure to what Rutter terms 'positive opportunities' in order to open up their horizons (Gilligan, 1999; Rutter, 1999). As many 'at-risk' young people have limited opportunities, and disadvantaged and impoverished backgrounds, set against other negative experiences, it may be even more important to provide them with opportunities in the community which may help to furnish them with 'protective factors'. In their study looking at the importance of the family, school and community, Howard and Johnson (2000) found that the young people themselves mentioned the importance of community groups, such as the Scouts, and of the skills they learnt there. In this research study, it was found that by the young people being motivated and wanting to be with the horses, they in turn developed new skills and an interest which could lead to further future opportunities in their lives.

Wayne spoke about how overcoming his fears with the horses at The Yard had enabled him to participate in a group outing which he had not had the confidence to do previously, as illustrated by the following extract from fieldnotes:

> Wayne proudly tells us that he went riding at another stable and had a canter. He is clearly really pleased and proud about having achieved this. I ask if he had gone out riding with the PRU [pupil referral unit] before but he says he always chose other things, or refused to do anything, previously.
>
> (fieldnotes)

98 *Equine-Assisted Therapy and Learning with At-Risk Young People*

Drawing on Rutter, Giller and Hagell's (1998) suggestion that protective mechanisms are linked to developing resilience, Ungar, Dumond and Mcdonald (2005) connect these to outdoor adventure programmes for at-risk young people. They suggest that participating in challenging, outdoor experiences offer similar protective mechanisms, such as promoting 'self-esteem and self-efficacy through experiences coping successfully with stress', and providing opportunities where 'hopefulness may replace feelings of helplessness' (Ungar, Dumond & Mcdonald, 2005: 325).

Hayden (2005) found that a major theme in her research with 'at-risk' young people participating in EFP was that of 'positive experiences'. She argues that through the young people having positive experiences with horses, this leads them to having greater motivation, participation and 'treatment compliance' (Hayden, 2005: 105). In turn this can lead to the opening up of the 'positive opportunities' that Rutter (1999) speaks of. By having gained the confidence to join in on a trip to a new riding stable, which previously he did not have the confidence to do, Wayne was able to have his first canter. This is something which most riders will describe as very exhilarating. Certainly cantering and galloping at speed for the first time gives a huge sense of achievement and heightened confidence in riding ability (if it is achieved without mishap, of course). Another aspect of the 'opening up of positive opportunities' is that of having horizon's opened up by having learnt and mastered new skills. By gaining confidence and learning a new skill, horizons may be opened up where previously they were limited. The participants Emma, Lucy, Kelly and Wayne all either expressed interest or became actively involved in further activities or studies with horses and/or other animals outside The Yard. In addition, the foster carer Angie explained how another boy she had on a temporary foster placement and who had attended The Yard during the study period had gone on to help out at a local stables near his new foster home, saying:

> He couldn't wait to get out to you, 'when are we getting there, when are we getting there', he *loved* it [he] did. In fact where he went after he left us he went somewhere where he was going up to the stables nearby every week, it was helping out and it really did him a lot of good...
>
> (Angie, foster carer)

During one particularly challenging TH session the participant, Kelly, was informed by her social worker that she was to be moving foster placements the following day with no prior notice. This was an extremely upsetting experience for Kelly and was an example of the additional trauma and anxiety that many young people in care are subject to. The section below provides an example of how it was hoped that through Kelly having the opportunity to discuss potential 'positive opportunities' during this, her final TH session, she was provided with some possibilities of a positive future:

Kelly's last session: 'I didn't know you could do that.'

Kelly arrived at The Yard looking quite down. She was quiet and withdrawn. After the RSW [registered social worker] had left she explained that she felt quite sick and had had very little to eat so didn't think she was up to doing much. However, when we gave her the option she didn't want us to call the staff to pick her up, but being mindful that it was sunny, Deana suggested that we take a chair into the barn and she could just sit and rest with the horses in the shade whilst we groomed the horses. We quietly groomed the horses in the cool barn, the birds were singing and it was very peaceful. We didn't put any pressure on Kelly to speak but gradually she began to join in, although she looked preoccupied and clearly didn't want to speak about what was on her mind. However, it didn't take much persuasion from us for Kelly to agree to go out for a ride. Around the horses Kelly seems to be able to shake off any heaviness that she arrives with. As soon as Kelly was on Ruby's back she noticeably relaxed and opened up to us. She spoke of her deep love of horses, her past experiences with them in relation to her childhood and early friendships and her hopes for the future, how one day she would love to have a horse of her own. Donna [horse-handler] told her of her own experience of being a working pupil at a stables, learning horse care and riding, living with other young people in a shared house, sharing their passion of horses, lessons, mucking out stables etc. Kelly said she would love to do this herself when she leaves school next year, telling us 'I didn't know you could do that'.

(fieldnotes and therapy sessional notes)

A further example of how experiences with the horses enabled the opening up of 'positive opportunities' was provided by another participant who ceased attending The Yard during the research study. We were later informed by this young person's social worker that he had been accepted at a local agricultural college to undertake an entry-level qualification in

100   *Equine-Assisted Therapy and Learning with At-Risk Young People*

equine studies. If it is accepted that educational achievement is a good thing, this would certainly seem to be a remarkable achievement for many of the young people who attended The Yard, who often possessed limited academic ability and even less educational motivation.

Challenge, motivation and learning

One of the themes to emerge from the data was how some of the young people saw the horses as a challenge. This was expressed in different ways, such as Wayne saying that he liked Jason 'because he is big!' Lucy talked about riding horses, explaining: 'I prefer horses who are more forward going.' In equestrian terms, 'forward going' means a horse that can be quite lively and requires a more experienced rider. When Lucy was asked why this was she replied: 'I think it's because I like horses who are a bit more of a challenge.'

To successfully work with and manage horses that are 'challenging', it is necessary to learn about horse psychology and be able to control and manage one's own behaviour and body language in order to communicate with the horse in a way that it can understand. Taking instruction and learning in the formal school environment had proved difficult for some of the participants who attended The Yard, a number attending PRUs and others having a Statement of SEN. However, it seemed that the challenge of wanting to be successful with the horses provided a degree of motivation to learn and to be able to overcome difficulties in taking instruction for some of the participants, especially Wayne, Minimax, Cinderella and Emma.

Emma's mother, Linda, explained this in terms of Emma's behaviour and how she had had to learn to modify her behaviour in order to work successfully with horses. In the following interview extract, Linda elaborates:

> I suppose in a way she has to understand some respect with horses because they're big. And you know, she just can't push them around and boss them around like she can sometimes with people of her own age, you know. She likes to think she's strong and she can control things with strength and bullying people and things like that. And I think probably with a horse, because the horse is big enough and it can, you know, if she was to... fight with the horse, the horse will win. I think there's an element of that. They're a challenge for her, you know. She has to be a bit more creative in how she wins whatever battles she has going on in her mind. Um, and so I think

Self-Confidence & Opening Up Positive Experiences 101

probably it's been a learning experience for her. They've the strength so they take her out of her [comfort zone] you know. Sometimes you've got to think differently and that has probably made her think sometimes.

(Linda, Emma's adoptive mother)

During the time she attended The Yard, Emma started weekly work experience at both a local equestrian college and an RDA centre as part of her school curriculum. This was because it was recognised that this was the most effective way for Emma to gain educational and socially beneficial opportunities. Emma said that one of the things she liked about horses and was part of the reason she was especially drawn to Jason, who could be quite a difficult horse, was because they were 'a challenge'.

For Emma it seemed that horses opened up possibilities and horizons. Linda explained how it was only through horses that she would participate in any school work. Emma had a statement of special educational needs (SEN) due to her disruptive behaviour at school and saw an educational psychologist. She was very open about her behaviour, saying: 'I do things like throw my food if I don't like it and stuff like that.' When Emma was asked what made her feel the need to do these things, she replied: 'I don't know, I just can't help myself.' Later though she did seem to offer some insight into this behaviour whilst talking about the similarities between Jason (the horse) and herself. This time she explained: 'I get like *totally* bored in school...that's why I'm really naughty to the teachers and wind them up...' In an interview with Linda when asked to expand on how she felt that horses had been a learning experience for Emma, she replied:

I think what has helped her, about horses, is that she understands that if she wants to learn more about horses that she has got to stop, sort of putting up this mental fight, um, about getting information, you know. That she has got to accept that some people do have more information. And I think she has learnt that now and that has been a huge step for her, and I think it's only because she is so motivated by horses that she's been able to do that.

(Linda, Emma's adoptive mother)

Reasons why horses are especially motivating despite the challenges they possess, due not least to their large size, have been proposed to be multifaceted. Authors talk about the challenges of working with horses due to their power, size and inherent wildness, but also

102   *Equine-Assisted Therapy and Learning with At-Risk Young People*

their vulnerability, curiosity and willingness to co-operate and be with humans (Karol, 2007; McCormick & McCormick, 1997; Vidrine, Owen-Smith & Faulkner, 2002).

The therapist Deana explained that, for her, the combination of the size and power of the horse together with its willingness to co-operate with humans created an almost magical experience:

> I think the thing about the horse is her size or his size, and there's something really special for me around that kind of that bulk, the volume. And the power. But also my experience is that there's gentleness to them ... you know the horse's [being] phenomenally powerful and strong, but actually she's also very vulnerable. Um, and I think there's something incredible when you're working with a child or even just yourself where the horse kind of willingly co-operates with you. And I think, um, a lot of people are quite afraid of horses um, and because of past experience or just because they're so big and strong. And so when a horse co-operates with you willingly, um, there's something really magical I think, must be happening.
>
> (Deana, therapist)

Emotion regulation through experiences with horses

Motivation to be with the horses seemed to enable some participants to regulate their behaviour and emotions, known as 'emotion regulation' in the literature. There is a large body of literature that looks at the many variables involved in emotion regulation, such as temperament differences attributed to early childhood influences, neurobiological reasons and environmental factors (Southam-Gerow & Kendall, 2002; Thompson, 1994). It is claimed that both a too high and a too low EI (the temperament-based index of emotionality) leads to risk factors, such as behaviour labelled as antisocial, disruptive and problematic, the optimum being moderate regulation of emotion (Southam-Gerow & Kendall, 2002). Emotion regulation has been defined as the 'extrinsic and intrinsic processes responsible for monitoring, evaluating and modifying emotional reactions ... to accomplish one's goals' (Thompson, 1994: 27). In the bestselling book *Emotional Intelligence*, which draws on similar concepts, Daniel Goleman makes the claim that characteristics of emotional intelligence, such as possessing self-awareness, impulse control, empathy and social competence, are more important than IQ (Goleman, 1996).

Southam-Gerow and Kendall (2002) suggest that 'emotion understanding', such as recognising emotions in others and understanding and behaving appropriately towards others emotions, are linked to cognitive development. They propose that delayed or limited emotion understanding may therefore put young people at risk of psychopathological disorders as they have limited social functioning and coping skills. The same authors talk about how therapeutic interventions can be useful in helping at-risk youth develop coping strategies and learn that 'emotions, even extreme ones, are not permanent, are endurable, and are not harmful in and of themselves' (Southam-Gerow & Kendall, 2002: 209). Southam-Gerow and Kendall (2002) go on to suggest that treatments that expose the child to fearful situations, within a supportive environment, can help to overcome anxieties, giving an example of working with a ten-year-old boy with separation issues and a fear of being in public spaces. Using techniques similar to those employed in mindfulness therapies, TH and EAT/L, the boy was encouraged to pay attention to his body feelings and natural environment, with the authors reporting that 'his inner world calmed down and he felt less anxious' (Southam-Gerow & Kendall, 2002: 209).

Chapter 6 explores themes around how contact with the horses seemed to enable the young people to feel calmer, and looks at links to TH and mindfulness in terms of how being with horses facilitates similar body experiences as those taught in mindfulness practices. It seems that there are parallels with the emotion-regulation literature and some of the themes that emerged from the study. By learning about horse psychology and being motivated to adapt and change their behaviour in order to work effectively with the horses, the young people were then able to regulate their emotions, which could lead to increased self-awareness as suggested as being important for successful development. Further examples of this are demonstrated by Cinderella in both the examples of her session when she found difficulty catching and haltering Duchess and with being able to control her emotions in order to help Leo when he was suffering. Lucy too demonstrated how she gained awareness of how her behaviour and emotions could affect the horse and how she learnt how to adapt this and, furthermore, translate it to help her with other areas of her life.

Conclusion

It has been demonstrated that by participating in what were perceived as challenging activities with horses the young people were able to gain benefits which are considered 'protective factors' in the literature on risk and resilience. It is claimed that young people understood as being 'at risk', such as those who were referred to The Yard, can be at greater risk of negative life outcomes, but that certain 'protective factors' can help to provide resilience to overcome this. Some of these protective factors are suggested to be self-confidence and possessing self-esteem, which in turn furnish a greater sense of self-efficacy, of being able to have some sense of control over your actions, environment and future. In addition it is claimed that being able to regulate emotions and behaviour, and to feel empathy with others, is important in terms of forming and building successful relationships and opening up 'positive opportunities'. It would appear through the descriptions provided in this chapter that participating in TH can help young people to gain a sense of achievement and accomplishment by having successful experiences with the horses, which can build self-confidence and self-esteem, increase mastery experiences, develop empathy and ultimately open up positive opportunities. In Chapter 4, related themes are explored: of how building up relationships, trust and attachments to the horses appeared to be linked to facilitating the young people's ability to participate in activities which then enabled them to gain additional benefits, such as the development of empathy, which is suggested as being important to resilience and healthy development.

4
Developing Attachments, Empathy and Trust through Relationships with Horses

Introduction

A major theme to emerge during the course of the research was to do with how the young people developed very meaningful relationships and attachments to the horses. The participants showed their attachments to the horses in different ways, as did the horses to the young people; it definitely appeared to be a two-way relationship in many cases. A number of authors speak of how many horses seem to have an innate sensitivity to young people and adults experiencing emotional distress in some way, appearing to demonstrate extraordinary tolerance and understanding (Isaacson, 2009; Kohanov, 2001, 2005; McCormick & McCormick, 1997; Meinersmann, Bradberry & Bright Roberts, 2008; Yorke, Adams & Coady, 2008). In many cases it seemed that the participants were able to display empathy with the horses, an emotion which is reported to be intrinsically linked to parental attachment styles (De Paul & Guibert, 2008) and therefore often lacking in young people who have suffered dysfunctional attachment patterns and abusive childhoods, such as many of the young people who were referred to The Yard. It is suggested that the development of empathy is imperative to 'the healthy emotional and social functioning of youths' (Thompson & Gullone, 2008: 123). Conversely, a lack of empathy is reported to be linked to antisocial and aggressive behaviour, and less success in later life due to difficulty in forming positive relationships (Lexman & Reeves, 2009; Thompson & Gullone, 2008). Waal (2009) argues that empathy and cooperation have been crucial to human survival and are more important than selfishness and competition in terms of evolution. Horses live in cooperative social structures, and it was interesting that many of the young people who attended The Yard seemed to identify and empathise with the horses and the different situations and challenges that they faced more so than they were reported to do with other people. Some of the participants seemed to choose horses that they felt needed them on some level, such as the participant Cinderella, who became very attached to the old pony Leo who suffered from health problems and required additional care and attention. How this appeared to be linked to themes of empathy is explored further on in this chapter.

These attachments to the horses seemed to work on a number of levels and to have different meanings for different young people. For some it appeared to be a way in which they could express affection and nurturing. For others there seemed to be identification with the horse's temperament, characteristics and background which corresponded with their attachments to them. Additional insights related to these themes are examined in more depth in Chapter 5 on the horse and the therapeutic relationship. A number of participants spoke of the horses as confidants – of how the horses understood them and of how they could trust them and 'tell them your secrets'. This idea, of animal as confidant, is something that has been noted by a number of authors in the AAT literature (Fine, 2000; Levinson, 1969; Melson, 2001; Myers, 2007; Turner, 2005).

The theme of being able to express affection and nurturing had an unexpected gender element as it was perhaps especially noticeable with some of the teenage boys, with their carers often expressing surprise at their openness in displaying affection to the horses. Some of the carers and social workers commented on how these boys found it difficult to display affection and show their feelings with people but would seem to be able to do this with the horses on some level. Both Minimax and Wayne would frequently spend time hugging and stroking the mare Ruby and talking to her quietly, in soft gentle tones. They would often disengage from what we were doing in a TH session and almost go into a trance, in their own world with the horse, and we would respect this time of what would appear to be a very healing experience for them. The value of touch and physical contact, and their relationships with horses by participants recovering from trauma, has been discussed by Yorke, Adams and Coady (2008). They suggest that 'the importance to healing of closeness, touch and physical contact between horse and rider' is an important element, one that is fraught with difficulty in the therapist–client relationship and which horses can perhaps provide without these ethical complexities (Yorke, Adams & Coady, 2008: 25).

Animals and attachments

Whilst the literature on the relationships between horses and children and possible links to attachment theory and empathy is still in its infancy, there is a more robust literature surrounding the field of AAT, as introduced in Chapter 1. Numerous authors speak of the therapeutic importance of animals in helping both adults and young people to experience healthy attachments to another living being (Beck & Katcher, 1996; Beck & Madresh, 2008; Corson & Corson, 1980; Hart, 2000; Levinson, 1972; Melson, 1990, 2001; Netting, Wilson & New, 1987; Parish-Plass, 2008; Rew, 2000; Siegel, 1993; Walsh, 2009; Wells, 2009). Melson proposes that animals provide much opportunity for children to experience healthy attachments and to explore transitional experiences through animals and play, mentioning Freudian and Jungian theories of animal symbolism and suggesting that animals are useful because 'children readily access animals as material in the development of a sense of self' (Melson, 2001: 20). She goes on to express the view that animals provide a healing role for children by providing companionship, emotional support and an opportunity to experience love and affection, in addition to their role as 'social lubricant', initially suggested by Levinson (1969). In her research with children and their pets, Melson suggests that, for some children, animals are more important to them than the people in their lives and that, 'for many children...pets are more likely to be part of growing up than are siblings or fathers' (Melson, 2001: 34). Tedeschi, Fitchett and Molidor support the view that animals are important for young people as they 'can provide opportunities for attachment and nurturing of others' (2005: 65). They suggest that animals can be especially valuable for at-risk young people as they 'can uniquely fill a combination of emotional needs, sometimes substituting for an absence of human attachment' (Tedeschi, Fitchett & Molidor, 2005: 65). In her AAT work with children with insecure attachments, Parish-Plass (2008) makes reference to object relations theory and how children can project issues onto animals in order to work through trauma. In addition, she suggests that AAT provides an opportunity for change of the internal working model that Bowlby refers to (Parish-Plass, 2008).

This acknowledged value of the attachment between a young person and a pet can be applied equally to a horse. Karol, an equine-assisted psychotherapist, speaks of how one of her clients, a young girl, described the therapy horse she was working with as her best friend (Karol, 2007). In their study looking at the therapeutic value of horse–human bonding

108 *Equine-Assisted Therapy and Learning with At-Risk Young People*

with six participants recovering from trauma, Yorke, Adams and Coady (2008) conclude that

> The results of this study suggest that the relationships participants had with their horses contributed significantly to their healing from trauma.
>
> (Yorke, Adams & Coady, 2008: 25)

They go on to suggest that the horses provide a parallel to the qualities required in a successful psychotherapeutic relationship, such as rapport, respect, warmth and acceptance (Yorke, Adams & Coady, 2008). The theme of how horses can perhaps provide a bridge to helping to build up a successful healing relationship between patient and therapist are explored in more depth in Chapter 5 on the horse and the therapeutic relationship. This chapter concentrates on looking at how the young people expressed their attachments to the horses and of associated observations arising through these relationships.

Nurture, 'holding', touch and empathy

The foster care company that initially funded the TH programme had specifically sought to recruit foster carers who lived in rural areas and had animals and/or a rural lifestyle, believing in the benefits of this for children in foster care. In an interview, Angie, Minimax's carer, explained that she had her own retired horse and of how Minimax had got very attached to this old mare. She talked of how Minimax would want to see the horse every day before school and was very affectionate towards her,

> and he hates it if he can't see [the horse] before he goes to school, wants to see [the horse], he goes in and cuddles her you see, and she's so big and strong, I suppose that he feels safe with her, I dunno, all the other horses love her because she's the boss.
>
> (Angie, foster carer)

Over the months that Minimax attended The Yard, he built up a strong relationship and attachment with Ruby. When he arrived he would always go straight to her and give her a long hug, as recorded in my following fieldnotes. Within foster care the area of tactile contact is fraught with difficulty due to the danger of allegations of inappropriate behaviour against the foster carer (Pemberton, 2010). With children

Developing Attachments, Empathy & Trust   109

who have previously been subject to inappropriate and abusive physical relationships, this is obviously a difficult area to manage as these children can display very dysfunctional and distorted behaviours around physical touch (Hughes, 2004). Hugging the horses was clearly very important for Minimax and was perhaps fulfilling a basic need for tactile contact (Frank & Frank, 1991; Yorke, Adams & Coady, 2008) that had previously been missing for him. The importance of touch and tactile contact with another living being has been demonstrated by the rather distressing, and controversial, experiments by Harlow on baby monkeys. His studies found that 'the infant monkeys preferred to cuddle with the softer, cloth surrogate mothers, regardless of the opportunity of receiving milk' (Levinson, 1980: 109).

The following extracts from fieldnotes with Minimax describe how he expressed his affection for Ruby:

> Once back in the yard Minimax headed straight to Ruby and put his arms around her neck, giving her a hug. He stayed hugging her for some time whilst we carried on with yard tasks around him until he eventually joined us in grooming the horses.
>
> (fieldnotes)

> Three months since Minimax last at the yard and he seems very happy to be back, immediately leaving Angie and going straight over to Ruby to give her a big hug, asking 'have you missed me Ruby?' He stands with her, his arms around her neck and his face buried into her long black mane for a good 5 minutes whilst we chat to Angie and catch up on how Minimax has been.
>
> (fieldnotes)

Wayne too would spend much of his time at the beginning and end of sessions hugging Ruby and gently stroking her face and neck in quiet contemplation. In one session where he finally got to ride her, we suggested some exercises for him to do on her back. These are essentially stretching and balancing exercises and often employed in vaulting lessons in order to encourage better balance and riding ability. Many of the young people enjoyed doing these and Wayne was no exception. In this particular session, we asked Wayne if he now trusted Ruby enough to feel confident to allow himself to lie backwards along her back. This is both a good stretching and balancing exercise, and one of trust in the horse and oneself. After an initial moment of hesitation Wayne willingly agreed and lay back along Ruby's back so that his head was almost touching her tail. The mare stood quietly and still as a rock whilst he did this, and Wayne remained in silence connected to the horse for a good few minutes. We too remained in silence, listening to the birds singing and the faraway mooing of a cow until eventually Wayne chose to rise to a sitting position again. Following this session the therapist, Deana, commented that this seemed to mirror a sort of 'holding' for Wayne on some level, and his YOT mentor, Peter, told us how Wayne appeared much calmer and more relaxed afterwards.

The clinical child psychologist Susan Brooks (2006) also suggests that this tactile element may be providing an arena for the child to be 'held' in this psychoanalytical sense as well as a physical one. She writes:

> Children who have been under-nurtured have a second chance to experience being wanted through touching and hugging large animals who can 'hold' them. Many of us have had a big grandmother who, when she pulled us to her and enveloped us in her big hug, allowed us to feel love and security. Animals, in many ways, can provide such a 'hug'; they can certainly give under-nurtured children a second chance to feel a sense of security and love.
>
> (Brooks, 2006: 202)

In addition, Scott (1980) writes about how touch is important in relation to attachment theory in terms of attaching to the therapist. It may be that for some of the young people who were demonstrating some sort of attachment to the horses that the horses were providing some of the aspects of healthy attachment that had been missing in their own early childhoods:

> tactile stimulation is a very important part of the attachment process. According to one line of clinical theory, the patient must become attached to the therapist if there is to be a successful outcome. This is especially important in the treatment of autism, where there is a derangement of the attachment process itself.
>
> (Scott, 1980: 139)

It is suggested that as well as the tactile opportunities offered by animals there are other nurturing qualities in caring for an animal which have an additional gender dimension. Levinson (1980) and Melson (2001) make the claim that caring for a pet can provide an acceptable outlet for boys to express and experience nurturing. They suggest that there are limited opportunities for boys to develop and display these important nurturing tendencies in Western societies. Melson describes this as follows:

> Caring for pets is the only outlet for nurturing others that is ubiquitously available for most boys in their homes and does not reflect a suspected diminution of masculine behaviour. This makes pet care a potential training ground for learning how to nurture another being who has different needs from one's own.
>
> (Melson, 2001: 58)

It appeared that through Wayne and Minimax building up a trusting relationship with Ruby and experiencing healthy tactile experiences with her, they were able to display affection and nurturing behaviours towards her. There may be connections to what some of the authors above suggest in terms of providing an acceptable outlet to experience some of the psychological elements which may have been missing from their childhoods, or are not considered acceptable masculine traits for boys to display in our society.

Horses helping to meet unmet needs and the development of empathy

Attachment theory talks of the concept of 'felt security', an internalised secure base, which, through normal healthy childhood development arises through the experience of being loved and receiving care sensitive to the physical and emotional needs of the individual child (Bowlby, 1988; Howe & Fearnley, 1999). By receiving 'good enough' parenting (Winnicott, 1965) and unconditional love the child can learn to love others and, in turn, go on to have healthy adult relationships. Conversely, where children instead experience neglect, abuse and inconsistent care, the ability to form healthy, trusting relationships is seriously compromised (Crittenden, 1988; Howe, 2005; Hughes, 2004). Destructive and difficult behaviours that are a consequence of living in survival mode can be a result and were often reported on the referral forms of the young people who attended The Yard. Whilst attachment theory is based on relationships between humans, some authors have suggested that the concept may be applied to the human–animal relationship (Beck & Madresh, 2008; Levinson, 1972; Melson, 1990, 2001, Parish-Plass, 2008; Tedeschi, Fitchett & Molidor, 2005; Walsh, 2009). This literature claims that for children and young people who have experienced unhealthy childhood backgrounds, contact

112 *Equine-Assisted Therapy and Learning with At-Risk Young People*

and forming relationships with animals can facilitate steps to forming healthy relationships with people (although see Crawford, Worsham & Swinehart, 2006, for a critique of whether human–animal attachments parallel traditional attachment theories).

Melson (2003) proposes that possessing empathy is a precursor to nurturance. Another study claims that there is a positive link between childhood pet-keeping and higher levels of empathy with both animals and people (Paul & Serpell, 1993). From a psychotherapeutic perspective it may be that by giving love and affection to the horses the young people who attended The Yard were, on some level, experiencing some of the nurturing, love and affection that had been missing in their childhoods. It is argued that attachment to a 'warm, sensitive, responsive and dependable' adult is required for healthy childhood development (Karen, 1998: 6). Where this has been missing, some authors suggest that building positive relationships and attachments with horses may fulfil some of these needs (Bowers & MacDonald, 2001; Gammage, 2008; Karol, 2007; McCormick & McCormick, 1997). Because horses are dependent on humans for their care and well-being, in the same way as a child is dependent on its parents or other caregivers, it may be that caring for the horses was a way for the young people to help another living being with the suggested associated benefits. Schwartz and Sendor (1999) claim that 'helping others helps oneself', and that participants who gave support to their peers showed pronounced improvements in confidence, self-awareness, self-esteem, depression and role-functioning. Additionally there would appear to be parallels with the importance of empathy in healthy development, and social and emotional functioning in later life (Thompson & Gullone, 2008). It is proposed that 'empathy is the single most desirable quality in nurturing parenting' (Bavolek, cited in De Paul & Guibert, 2008: 1064). Accordingly it is argued that being able to respond empathically towards their child is considered to be a main protective factor in successful parenting and child development, and, conversely, child neglect issues are a consequence if empathetic parenting is lacking (De Paul & Guibert, 2008). Lexman and Reeves describe empathy as 'an ability to put yourself in another person's shoes – and to act in a way that is sensitive to other people's perspectives' (2009: 17). They make reference to studies that support the belief that empathy develops alongside healthy attachment patterns with a positive carer in early childhood. They go on to make connections with the neurobiological literature which suggests that empathy is 'hardwired' through synapse connections formed as a result of responsive positive, empathic care. Other authors suggest that the opposite happens as a result of neglectful and abusive early childhood experience and that these children do not therefore develop empathy, with antisocial behaviours resulting (Gerhardt, 2004). Empathy is one of the characteristics that is vitally important in determining success in later life, due to its importance in relation to forming successful working and personal relationships, according to a study by the think tank Demos (Lexman & Reeves, 2009), and has links to the risk and resilience literature.

It may be that for young people who have been unable to develop empathy due to dysfunctional attachment patterns, and empathy not having been demonstrated to them by their adult caregivers, by building a relationship with an animal the beginnings of the development of empathy may be facilitated. Whilst research looking at the development of empathy in these groups is limited, a number of authors make the suggestion that animals can help children in general to develop empathy, 'interaffectivity' and 'moral development' (Levinson, 1980; Melson, 2001, 2003; Myers, 2007; Poresky, 1990). How the horses often appeared to give the young people the opportunity to give them love and affection was expressed through the way in which they behaved with them, often hugging them and telling them they loved them. When asked about her relationship with Leo (her favourite pony) and how she thought he liked her to behave with him, the participant Cinderella replied: 'Just giving him love...and...and they like it if you talk to them in a soft, low-toned voice.' Talking to the participant Emma about some of the things that she liked about being with the horses, she said: 'They need you to look after them so it makes you feel needed.' For Emma, the fact that she felt the horses depended on her seemed to give her a sense of responsibility and feeling of being needed. In her study with homeless youth, Rew (2000) found that many of the young people spoke of how attached they were to their dogs, who gave them companionship, and how, in turn, they felt responsible for their welfare. This also gave them a reason to engage in healthier behaviours as 'having a dog makes me feel like I gotta stay healthier so the dog's okay' (Rew, 2000: 129).

The relationship that Cinderella developed with the pony Leo appeared to contain some elements of empathy. From her very first session, she was immediately attracted to Leo, an older pony who sometimes suffered from a painful foot condition called laminitis, and was very concerned about his health, wanting to spend much of her time in TH sessions grooming and caring for him. For Cinderella the relationship that she built up with Leo seemed to be based on her wanting to give him love and attention, aspects that, according to her referral form and social worker, had been lacking from her childhood and subsequent fragmented life in care. We had been informed that Cinderella

114 *Equine-Assisted Therapy and Learning with At-Risk Young People*

had experienced numerous foster placements which frequently broke down. This was to be the case whilst she attended The Yard, moving foster placement with very little notice to another county during the summer that she came for TH sessions. In a questionnaire about their experiences of TH that some of the participants completed for the research, Cinderella wrote this about Leo:

> I love Leo the best because he's an old boy who needs loads of TLC. He loves being groomed and cuddled feeding and general attention and stuff and I feel I can give him love. Today when we measured the hard feeds out into the buckets for the horses I gave him a bit extra coz' he's special!
>
> (Cinderella, questionnaire response)

This theme of feeling that the horses needed them, and of being able to provide nurture to them in some way, was mirrored by another participant, Lucy, who switched her affections from the more independent mare, Ruby, to Timmy, a new pony to join The Yard. Timmy was a rather anxious pony who got very agitated at new situations and was especially terrified of donkeys. One day out hacking in the lanes on Timmy and Ruby, we came across a field containing two donkeys which caused Timmy to display great fear, trying to run away and prancing and snorting himself into an anxious sweat. However, with encouragement, Lucy was able to remain calm and composed, and help Timmy to overcome his fear and pass the donkeys. Later, when asked about her feelings towards the two horses following this incident, Lucy replied:

> I think I've got more of a connection with Timmy now really, it's like I know him better now and he sort of needs you more, it's like Ruby doesn't really need you, she's more sort of solid and sure of herself, like she doesn't really need anyone
>
> (Lucy, participant)

This incident also provides an illustration of how wanting to work effectively with the horses enabled some of the participants to modify their behaviour. In the case with Lucy and Timmy, Lucy had to overcome her own initial anxiety and fear of Timmy's potentially dangerous behaviour (running off down the road) in order to remain calm and confident, and to provide a role model to Timmy. A further example of this is provided in the following extract about Cinderella and her relationship with Leo.

Modifying behaviour through empathy

It seems significant that Cinderella chose Leo, who was the most fragile and defenceless pony in the herd, as her favourite pony at The Yard. Cinderella could be rather loud and dominant in her manner, both in her voice and body language, and had difficulty in respecting personal space boundaries. Her referral form stated that 'Cinderella can display aggressive and defiant behaviour', and we experienced a testing of boundaries on her first session when she tried to light a cigarette despite being only 15 years old and a clearly displayed 'No Smoking' sign. However, it seemed that her love of horses enabled her to overcome this defensive behaviour and, by choosing Leo, who needed lots of gentle and calm care and attention due to his age and health condition, Cinderella had to monitor and change her behaviour to look after him. The following extract from sessional and fieldnotes is of a difficult session where Cinderella had to overcome her initial wish to run away when we found Leo lying down in the field after he had suffered an acute attack of laminitis (a painful foot condition) and needed veterinary attention.

Leo, Cinderella and the vet

Quite difficult session today! Had both Cinderella and another girl together today as this girl has recently moved into [foster] placement with Cinderella. Cinderella not happy about this and the two refused to do anything together so we made the decision to split the two up this week, with me spending the session with Cinderella and Leo and Donna with the other girl and Ruby (although still within the boundaries of The Yard).

Cinderella arrived and went straight to field to see Leo as she didn't want to be anywhere near the other girl. However, before I reached them she came running back crying and saying that he couldn't move. It was obvious that he had got an acute attack of laminitis. I called the vet but Cinderella found this really difficult and said she couldn't bear to stay with Leo whilst he was treated and had an injection. She then threatened to run off. But when I asked her what she thought Leo would prefer and what would be best for him, she reluctantly agreed to stay. She was then really brave and caring, holding him while the vet gave him his injection, even though she obviously found this really hard and held her face away. Cinderella then helped me collect wood-shavings and make a big deep bed in the barn for

116 *Equine-Assisted Therapy and Learning with At-Risk Young People*

him so he would be comfortable. Once we got him in to the barn Leo lay down and she lay down on the shavings with him, cradling his head in her arms. She asked us to take some photos of the two of them together on her mobile phone which we did.

(fieldnotes and sessional notes)

From a psychotherapeutic interpretation, a number of theories may be applicable in this encounter. Cinderella certainly showed tremendous strength of character to overcome her initial, and usual, reaction to difficult situations, which according to her referral form, and from our experience, was to either run away or become defensive and possibly aggressive. If, through her early life experiences, Cinderella had not had her needs met together within a secure holding environment, she could have remained in a state of anxiety, possibly lasting until adulthood. Winnicott believed that the defenceless infant needs to be given a secure holding base in which to be psychologically able to develop a secure sense of self in the world (Winnicott, 1965). Having a healthy attachment to a loving, attuned caregiver within the early years has become accepted as fundamental in understanding human development (Bowlby, 1988; Howe & Fearnley, 1999; Hughes, 2004). If Cinderella had not developed secure attachments, had learnt that her needs were not met, and, possibly worse, experienced abusive, inconsistent care, it is likely that she would have needed to develop coping strategies of becoming defended (Winnicott, 1984) to protect herself in an attempt to 'ward off the sense of helplessness, deflation and dependency' (Almaas, 1988: 375). It could be argued that Cinderella had had to learn to survive by relying only on herself (Hughes, 2004).

The Yard drew on philosophical therapeutic concepts drawn from many traditions, including core process, person-centred and Buddhist psychotherapy, which have as a primary belief that the 'true self' wishes to be fundamentally healthy and is simply obscured by the 'protective false self' (Mearns & Thorne, 2000; Rogers, 1951; Sills, 2009; Winnicott, 1965). In the encounter between Cinderella and Leo outlined above, something in her relationship with Leo had enabled Cinderella to change her normal behaviour in this instance. Perhaps by beginning to develop an attachment to Leo (which could be interpreted as being a projection of herself) and, in turn feelings of care and empathy towards him (and herself), she was able to obscure her 'protective false self', temporarily at least, and allow her 'true self', which fundamentally wants to heal, to emerge. The hope from a psychotherapeutic analysis would be

Developing Attachments, Empathy & Trust 117

that this experience would enable Cinderella to learn that she was 'able to relate to others (and herself) in a more positive and loving manner' (Gammage, 2008: 5).

Nearly a year later, after Cinderella had moved foster placements and stopped coming for TH sessions, she found the work mobile telephone number in a leaflet about the TH programme and sent the following text:

> Hey Hannah it's [Cinderella] the one whose favourite horse is Leo how is he doing hope he's alright.
>
> (Cinderella, text message)

This communication served to reinforce the attachment, and ability to empathise, that Cinderella had formed to Leo, remembering him many months later. Melson suggests that an important aspect of attachment is 'the motivation of the attached person to maintain closeness and continue the relationship' (Melson, 1990: 93). She also goes on to suggest that, in the case of children's attachments to pets, a child may feel close to a pet without necessarily spending much time with it. It would appear to be the emotional tie which is an important element and perhaps involved in developing a healthy internal working model. Bretherton (1985) suggests that children form an internal working model of every attachment relationship, and that these are developmentally significant as they enable the child to access ideas and feelings about relationships even when the attachment object is physically absent. This internal working model can then be applied to other, similar relationships (Bretherton, 1985).

De Paul and Guibert cite studies which claim that 'a person is more likely to experience "empathy" when the observer experiences a psychological sense of merging with the observed' (2008: 1067). In Chapter 5 on the horse and the therapeutic relationship, it is shown how many of the participants at The Yard seemed to strongly identify with certain horses. It may be that elements of this 'merging' may be relevant in the development of empathy. In the case of Cinderella, a young person who had experienced dysfunctional attachment models in her childhood, it could be that by having experienced a healthy attachment and relationship with the pony Leo, demonstrated by thinking about him and wanting to know how he was some time into the future, she was enabled to form the beginning of an element of a healthy internal working model of attachment, together with the ability to form empathic responses towards another living being.

118 *Equine-Assisted Therapy and Learning with At-Risk Young People*

Relationship, safety and trust

The ways in which the participants built relationships with the horses were expressed in different ways and could be subtle and slow to develop, as in the case of Minimax and Ruby, and Emma and Jason, or instantaneous, as with Cinderella and Leo. The extract below provides an example of what appeared to be the beginnings of a relationship building between Minimax and Ruby, and of how Minimax interpreted Ruby's actions towards him. This section from fieldnotes was following a session where Minimax had been working with Ruby in the round pen leading her around a small obstacle course consisting of poles and cones. It was an interesting phenomenon, which we witnessed time and time again, that many of the young people would choose horses that mirrored 'issues' or behaviours that they identified with. We wondered if aspects of projection were also at play in the encounters, and this is explored in more depth in Chapter 5.

Minimax and Ruby: Identification and connection?

Minimax completed the task successfully although he could be rushed in his actions without thinking through the route first. This resulted in him getting his foot slightly trodden on by Ruby. However, this was also around the same time that Ruby was displaying some anxiety from being separated from her fieldmate Leo (they are very attached to each other and show distress when separated) and Minimax may have been affected by this as he became a little tearful.

We are aware from his referral form and from speaking to his foster carer, Angie, that Minimax experiences huge emotional loss and subsequent behaviour problems due to being separated from his brother who was in the process of being adopted. Being mindful of this we suggested to Minimax that we return Ruby to the yard to be reunited with Leo and put his foot in a cold bucket of water to prevent swelling. Sitting on the bench with his foot in the cold water Minimax remained a little quiet and tearful. We were aware that Ruby had only slightly trodden on Minimax's toes and there was obviously no damage to his foot but that Minimax was obviously expressing a need to be tearful for other reasons and we allowed him to do this with no pressure. During this time a remarkable thing occurred. Ruby left Leo and the other horses and purposefully walked over to us putting her head in the bucket of water to take a drink, splashing the water onto Minimax's leg with her nose in the process and making us

Developing Attachments, Empathy & Trust 119

all laugh! This immediately cheered Minimax up, especially when we pointed out that Ruby could have much more easily gone and drunk from the (nearer) water tank but had chosen to come over and take a drink from his bucket. He asked 'do you think she's saying sorry for treading on my toe?'. We replied that it did seem as though Ruby was initiating a connection with him in some way, even though we were not sure if horses understand sorry.

(fieldnotes)

At the risk of criticisms of anthropomorphism, it certainly appeared that this action from Ruby was outside her normal behaviour as the water tank she normally drank from was nearer to her and the other horses, and she had purposefully left the other horses to come back over to Minimax to put her head in the bucket that he was cooling his foot in. We wondered whether other unconscious, unknown factors were at play in this encounter and of the possible projections going on with Minimax choosing to work with a horse that experienced and displayed very similar 'separation issues' to himself. Minimax chose to work with a horse that appeared to mirror his own feelings, and this may have been why he identified with her. Indeed, at a later date, Minimax went on to explain that he felt Ruby 'understood him' because of their similar pasts. There are many examples of these unexplained actions and encounters between horses and people, and numerous authors speak of how they have experienced and witnessed profound healing from encounters with horses (Baker, 2004; Isaacson, 2009; Kohanov, 2001, 2005; Richards, 2006; Soren, 2001). Whilst these are clearly unknowable and unanswerable questions, it would seem that the fact that Minimax felt that Ruby was making a connection with him in some way, which he interpreted as an apology for her hurting him, was therapeutic and comforting to him in a way which was meaningful for him.

Like some of these other authors above have noted, one of the therapists at The Yard, Deana, talked of how she noticed that it would appear that some of the horses would behave differently, and sometimes in a maternal, caring way, according to the participants' needs. She explained:

For I know I'm sure, I mean, you know, I've seen Ruby when she...she's almost been like a different pony with different people. You know, sometimes she can be quite mischievous and adventurous and another time if she's working with someone who is more nervous would be gentler and calmer and ...and...more motherly, actually.

(Deana, therapist)

120 *Equine-Assisted Therapy and Learning with At-Risk Young People*

This sense that the horses appeared to be willing partners in the therapeutic process, and that these relationships and connections seemed to operate for both the participants and the horses, was also noted by the foster carer, Angie, who said that she believed that 'the horses seem to sense that a child is troubled'. Sally, a counsellor and EAT/L therapist who worked at The Yard in addition to another centre, explained how her horse would behave differently with her clients compared with experienced horse people,

> But somehow when she's working with these youngsters she knows something. I don't know what it is she knows, but she does. She knows enough to stand ground-tied with people all around her, clients all around her, feet being kicked up, being groomed at the same time, you know, other animals are about, and she doesn't move.
>
> (Sally, counsellor)

Some of the participants expressed this by speaking about how they felt the horses understood them, and how they both identified with each other, as the following section about Minimax and Ruby illustrates.

'She understands me': Trust and attachment

In a questionnaire about his experiences with the horses, Minimax expressed his attachment to Ruby as follows: 'I love Ruby because she's quiet and understands me.' In an informal interview with Minimax whilst sitting at the bench in the yard in the sun with the horses milling about, I asked him if he could explain what he meant by this, and if he could talk about his relationship with Ruby in more depth. Minimax answered that he believed that horses 'have to learn to trust you'. When he first started to form a relationship with Ruby, Minimax had been very interested in where Ruby had lived previously to coming to The Yard and what her life and treatment may have been like. He went on to explain:

> I think Ruby trusted me straight away 'coz she knew I had the same past as her... I think she could just sort of tell.
>
> (Minimax, participant)

An example of how this trust was reciprocal was made evident to us in a session with Minimax when we went into the fields to collect the horses and Ruby was lying down, relaxing. My extract from the fieldnotes that day reads:

Developing Attachments, Empathy & Trust 121

When we went out into the field to bring in Ruby and Jason, Ruby was lying down resting in the sun. Minimax went straight over and sat down on the grass next to her. He seemed very relaxed with her, as did Ruby with Minimax, remaining quite happy to remain lying down which is often unusual for horses as they are obviously extremely vulnerable lying down and will only remain so with someone they trust and feel secure with.

(fieldnotes)

Another participant, Kelly, spoke of how she felt that the mare Ruby 'understands me' and that the reason she felt drawn to wanting to work with this particular horse was because 'she is calm and it's like I feel a sort of connection with her'. Kelly's therapist, Deana, who led these TH sessions with Kelly, had explained how closed Kelly had been in the previous six months prior to her attending The Yard. It was certainly noticeable how quiet and unresponsive Kelly had been during her first few sessions, it being very difficult to engage with her. However, over the next few sessions she did begin to open up and this appeared to coincide with her beginning to form a relationship with Ruby. In conversation whilst out riding Ruby in the lanes one day, Kelly began to relax and talk more about her feelings towards the mare. She said that she felt that horses 'can just sense your moods, sort of like . . . it's like they can just tell how you're feeling'.

'They won't tell your secrets': Horses as confidants

Another way in which the participants appeared to find their relationships with the horses valuable was as confidants who they could talk to and tell secrets. This is an area which has been noted in AAT, beginning with the psychologist Levinson, who introduced pets into his psychotherapeutic work with children as he believed that they could offer opportunities as confidants for the young people, amongst many other benefits (Levinson, 1969). Chardonnens points to studies which claim that children 'talk more easily with their animals than with their therapists' and about how the animal can be considered as a catalyst for social interactions (Chardonnens, 2009: 323).

The therapist Deana, talking about one of the participants, Freya, with whom she worked both at The Yard and at her residential placement, explained that that, when she first met Freya,

She was very defensive and, um, but we had had several conversations around dogs and horses and she, um, I got the

122 *Equine-Assisted Therapy and Learning with At-Risk Young People*

impression from her that she liked animals more than she liked human beings.

(Deana, therapist)

Bowers and MacDonald (2001) suggest that animals can meet some of the needs of children who lack a secure attachment base. They claim that children who have experienced abuse and trauma are more likely to turn to animals when they are stressed, as animals can offer some of the unconditional love and caring missing from these children's lives. In their study on equine-facilitated psychotherapy with 'at-risk adolescents', Bowers and MacDonald propose how the horse can be especially effective in developing alternative secure attachments due to its unique characteristics. They propose that the hands-on interaction between horse and child fosters opportunities for the child to learn to trust another living being as they learn how the horse begins to trust them once they are able to modify their behaviour to interact with it in a positive way (Bowers & MacDonald, 2001).

Many of the participants at The Yard talked of how they felt it was also important for the horses to trust them. How the young people understood this is demonstrated during one early session with Cinderella, who was experiencing difficulty catching the mare Duchess in the field (see Chapter 6 for a more detailed description of this session). This was due to Cinderella's body language being rather dominating and threatening to Duchess, who merely walked out of reach of Cinderella when she tried to approach her. After initially giving up and throwing the headcollar on the ground in anger, Cinderella finally calmed down and sat down in the field to work out what to do next. We took this opportunity to ask her what different approaches might help Duchess to feel more secure and encourage her to want to be caught. Expecting her to suggest returning to the yard to fetch food to entice her with, we were surprised when instead Cinderella answered: 'Well, probably getting to know me a bit more first, you know, so she knows she can trust me.'

Some authors offer the suggestion that the repetitive rhythmic pattern of horse-riding offers a connection to the horse that is healing in some way (Game, 2001), others that it can open up receptors in the brain which are linked to calmness and learning (Isaacson, 2009; Moorhead, 2010). Evans and Franklin talk of how the rhythm of riding, specifically dressage, 'links emotions and motions' to produce the ultimate 'moments of floating harmony which take them outside a ground-bound existence' (Evans & Franklin, 2010: 176). Further ongoing research is looking at how being with horses can reduce heart rates similarly to the way that stroking dogs has been claimed to reduce blood pressure (Friedmann *et al.*, 1983; Gehrke, 2006; Mistral, 2007). Although some of these theories have yet to be empirically proved, it was found with astonishing regularity that participants would start to talk and open up whilst out hacking (riding in the countryside). On one occasion when out riding with Emma, she began talking about the other animals she had at home, which included a cockerel, Charlie, and a new puppy. She likened them to family members, saying that 'Timmy [the pony] is like the big brother, [the cockerel] is like middle brother and [the puppy] is like the little sister'. When asked if she meant that the animals were like family members to her, she replied: 'no, animals are *better* than family because they don't tell anyone your secrets'.

This element, of animals being better than people because they will keep secrets, was replicated by Cinderella. In an informal interview during one session, she too talked of how the horses 'won't tell anyone your secrets' and wrote the following on a questionnaire:

> you can just talk to the horses and they won't answer back. They won't tell you're secrets and when you see their ears twitch you know they're listening.
>
> (Cinderella, participant)

It would appear that the participants felt safe enough with animals and horses to open up to and tell them their secrets. Safety and trust in the therapeutic relationship is argued to be fundamental to enabling change to happen,

> Creating this trust is the main initial task of psychotherapy. Indeed, this is the crucial task of psychotherapy, because once it is established, the patient can then pursue what he needs to, knowing he is not alone with it.
>
> (Sedgewick, 2001: 95)

By trusting and feeling safe with the horses it could be argued that this could be a starting point to developing a trusting relationship with others in the therapeutic environment. Participants expressed this theme in different ways. Lucy explained how she felt safe with one of the horses, Duchess, despite her large size, partly because of her older age. The element of the horse having a motherly instinct also appeared to be an important factor to Lucy, who felt that Duchess wanted to protect her in some way:

> Yeah, but it don't seem to matter [that she was so large] because she was such a nice ride, a nice horse to ride. Felt right to ride. Also,

she was older again, so you felt really safe, that was the other thing. It was that kind of feeling that she'd been around the block and she knew what was there, and because she had had a foal, her motherly instinct was quite strong, so she wanted to kind of protect you from the danger...to me ...of going around the block.

(Lucy, participant)

Davies, Winter and Cicchetti (2006), writing about emotional security theory in the context of children who have experienced intrafamily domestic violence, claim that 'within the hierarchy of human goals, protection, safety and security are among the most salient and important' (Davies, Winter & Cicchetti, 2006: 709). It may be that the horses offered the participants a sense of safety and security which had been missing from their lives previously.

Conclusion

This chapter explores how many of the young people appeared to develop strong and meaningful relationships and attachments to the horses. In turn, themes of nurture, trust and safety, and of being able to empathise with the horses, are described. The ability to empathise is considered to have importance in terms of being able to build positive and successful relationships, and is related to attachment, and risk and resilience theories. By developing attachments to the horses it has been suggested that young people who have experienced 'disorganised' attachments and abusive, neglectful and dysfunctional childhoods may be able to recreate and receive some of the nurturing and unconditional elements missing from their lives through meaningful relationships with horses. In addition, it has been demonstrated that horses can provide an alternative healthy tactile relationship, argued to be an important element in childhood developmental experience, and that they can provide opportunities for young people to identify with their individual experiences and characters.

Chapter 5, on the horse and the therapeutic relationship, expands on how some of these themes, of developing a feeling of safety and trust with the horses, can perhaps transfer to establishing a therapeutic relationship between the young person and the adult practitioners and therapists. It also explore themes of projection and identification, and other related psychotherapeutic processes.

Figure 1 Making friends: having a scratch

Figure 2 Horse care: plaiting a mane

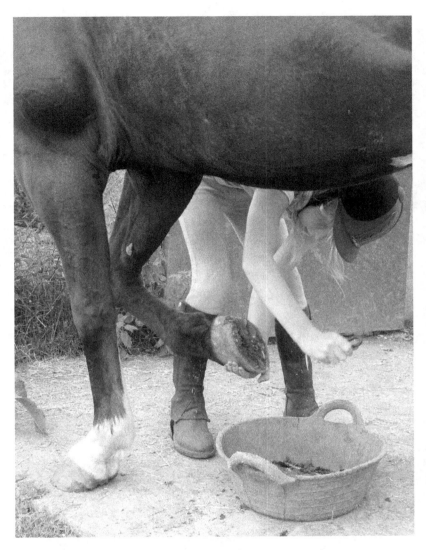

Figure 3 Horse care: picking out hooves

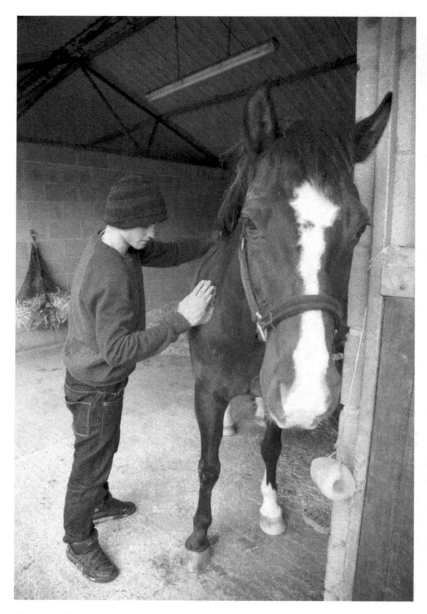

Figure 4 Horse care: grooming

Figure 5 Shetlands like hugs too

Figure 6 Group 'invisible leading' session

Figure 7 Exercises in the round pen: walking in rhythm

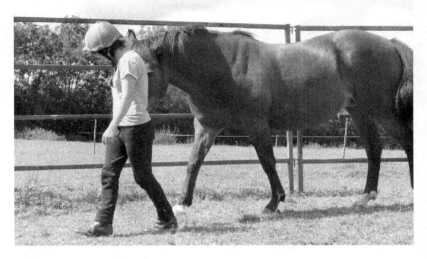

Figure 8 Exercises in the round pen: follow the leader

Figure 9 Exercises in the round pen: join-up

Figure 10 Horse agility: negotiating obstacles

Figure 11 Horse agility: building trust and confidence

Figure 12 Horse agility: crossing bridges

Figure 13 Horse agility: investigating the box

Figure 14 Horse agility: testing the box

Figure 15 Horse agility: success!

Figure 16 Exercises on horseback: stretching and balance

Figure 17 Exercises on horseback: balance and confidence

Figure 18 Equine-assisted education session

Figure 19 Equine-assisted education session: Jasper the horse helping with worksheet

Figure 20   Taking some time out together

Figure 21   Good friends

Figure 22 Getting to know each other: hand grooming

5
The Horse, the Therapeutic Relationship and Other Psychotherapeutic Insights

Introduction

This chapter looks at a number of themes related to the psychotherapeutic literature. It explores how the relationships that the young people developed with the horses could be seen to have some parallels with the therapeutic relationship between client and therapist, and also at how building up relationships with the horses seemed to correspond with the young people building up relationships with people: the horse appeared to be a facilitator to the young people being able to connect to others. Because the adults' views were more strongly represented in this area, it has drawn more heavily on data from the adults who contributed to the study than in the previous chapters. Another theme to emerge which is linked to how the young people related to the horses was of how they identified with them on different levels. Psychotherapeutic themes of identification, analogy, introjections and projection are also relevant and are explored in relation to the participants' interactions with the horses in addition to the therapists' views.

A central tenet within client- or person-centred and humanistic therapies is the quality of the therapeutic relationship. Jung stated that 'a close rapport is needed' (Jung, 1983: 166) whilst within Rogerian therapy it is referred to as 'congruence' (Rogers, 1967, 1992). The therapist Carl Rogers, from whose work client-centred therapy has evolved, claimed that 'significant positive personality change does not occur except in a relationship' (Rogers, 1992: 828). Within this framework it is understood that in order for the therapeutic process to be effective, a trusting 'therapeutic' or 'helping alliance' is imperative (Frank & Frank, 1991; Mearns, 1994). Rogers suggests that one of the most important qualities for an effective counsellor or therapist to possess is a great self-awareness of their own feelings and emotions in order to be able to meet the client on a 'person-to-person basis'. He describes this as '*being* himself' (Rogers, 1967: 108). Within EAT/L, many authors talk of the horse being an effective therapist in its own right as it knows nothing else but to be itself: a horse does not understand how to put on a false self or behave in any other way but with congruence because it is an instinctive animal (Frame, 2006; Karol, 2007; McCormick & McCormick, 1997; Meinersmann, Bradberry & Roberts, 2008). It is argued that through beginning to build rapport and feeling safe in a relationship with a horse, the young person can begin to feel safe in the therapeutic relationship, and this can then lead to positive change and learning (Bowers & MacDonald, 2001; Brooks, 2006; Karol, 2007; Meinersmann, Bradberry & Roberts, 2008; Yorke, Adams & Coady, 2008).

Other concepts, such as Rogers' 'unconditional positive regard' (Rogers, 1992) and the Winnecottian safe, or holding, space (Winnicott, 1965), could also be argued to be applicable to the horse–human therapeutic relationship. Because relationships with adults have, more often than not, been fraught with confusion and betrayal for many young people in care, such as the participants at The Yard, the importance of a non-judgemental relationship with another living being is perhaps even more imperative, something also suggested by Levinson (1969, 1980). Writing about the role of therapy dogs with children, Lori Friesen states that 'children seem to perceive them as non-judgemental participants who are outside of the complications and expectations of human relationships' (Friesen, 2010: 261). In addition to a non-judgemental attitude, other characteristics of the therapeutic relationship are suggested by Yorke, Adams and Coady, who suggest that

> a good helping relationship, characterised by mutual liking, respect, rapport, trust, warmth, acceptance, and collaboration, is the most powerful predictive factor for successful client outcome.
>
> (2008: 17)

It is claimed that many of the above qualities can be provided by the horse (Bowers & MacDonald, 2001; Brooks, 2006; Karol, 2007; Vidrine, Owen-Smith & Faulkner, 2002; Yorke, Adams & Coady, 2008). Certainly it will be seen that the words that many of the participants used to describe their relationships with the horses contained words such as 'trust', 'respect', 'calm' and 'safe'. An example of this is Freya's description of why she was especially drawn to the mare Ruby, saying: 'I like her [Ruby] because she is kind...and calm'. This suggests that the young people were perhaps experiencing some of the indicators provided above by Yorke, Adams and Coady (2008) of a successful therapeutic relationship.

By virtue of starting to feel safe with the horse and experiencing some of the above factors, which are claimed to be crucial in therapy, the young person may be able to start to express how they are feeling, and consequently the therapeutic healing process may begin. Many of the young people seemed to identify with certain horses and want to explore how the horses felt about situations such as partings, moving to a new home, meeting new horses and past experiences. This was something also found by Trotter *et al.* who write: 'participants were often instinctively drawn to a horse that had characteristics similar to their own' (2008: 263), and was a theme found in a previous riding therapy study with adults (Burgon, 2003). It may be that this was a way for the young people to explore and process certain issues that were relevant to them, but in a way that was one step removed and so, perhaps, safer.

Building relationships: Horses and 'unconditional positive regard'

As has been discussed previously, being with the horses seemed to help to provide a feeling of safety to some of the participants. The therapist Deana described how she felt that this then transferred to her being able to develop a therapeutic relationship with the participant Kelly. Deana described how Kelly called her by her name for the first time during one of her early sessions at The Yard. Deana had had been seeing Kelly for over six months previously at the residential facility where she had been placed by social services, and had found it very difficult to engage with her. At her first few sessions, Kelly was very quiet and reserved, but as she grew in confidence and started to build up a relationship with Ruby, she became more communicative, as demonstrated in the following section from fieldnotes.

Kelly and Ruby: round pen work

Kelly – very sensitive and quiet manner with Ruby today, showing patience and perseverance with her as Ruby was initially reluctant to leave the yard. Then, up in the round pen, she was distracted by some geldings in a nearby field – to which we laughed together as Kelly said 'I know some girls like that!' Initially Kelly had a tendency to be unconfident and ineffective in her aids [instructions] and body language. However, she soon discovered that by remaining calm but

128　*Equine-Assisted Therapy and Learning with At-Risk Young People*

having confident body language and being persistent, Ruby began to listen to her and lead and halt when Kelly asked with very subtle aids. Kelly was clearly pleased with the positive feedback this gave her and became more communicative and confident in her manner with us.

(fieldnotes)

In our debrief meeting following the TH session, Deana described how it was in this session that Kelly spoke to her by name for the first time. Later in an interview with Deana when asked to expand on this, she said:

and it was very, very, difficult to engage with her. She was very defensive...and, well, yeah, I mean, I'd been working with her for six months and she hadn't actually called me by my name until we came here, and we'd been here for only a few sessions and then she called me by my name and...and I think that for me just...um, I saw such a contrast in her. I think that was...that was what was so clear for me was the time I'd been working with her, if that was the only side of her that I'd ever seen, I would have said, you know, she's very kind of um, I might use words like hostile or defensive or difficult to engage, but here, I mean, she was like another child. She was like another person. Much softer, more open, more trusting, more willing to engage with me, and I could see a difference in her face when she was here to when I was working with her in the other place. And...yes, just more kind of open, trusting.

(Deana, therapist)

Rogers talks of 'unconditional positive regard' and empathy as being necessary conditions to facilitate therapeutic change. In terms of unconditional positive regard, he says that this means that there are no conditions attached to acceptance by the therapist, and, for empathy, that the therapist has an accurate, empathic awareness of the client's experience (Rogers, 1967, 1992). Both of these principles have been argued to be contained within the relationship possible with the horse and other animals (Bowers & MacDonald, 2001; Chardonnens, 2009; Fine, 2000; Vidrine, Owen-Smith & Faulkner, 2002). The horse has no preconceived judgements of the person before it and, additionally, some authors argue that horses possess an innate understanding and empathic behaviour towards many people suffering from social/emotional difficulties of some kind (Isaacson, 2009; Kohanov, 2001; McCormick & McCormick, 1997; Yorke, Adams & Coady, 2008). Whilst issues of

Psychotherapeutic Insights   129

anthropomorphism may be levied at these claims, it could be seen previously with the encounter between Minimax and Ruby after she trod on his toe that it was meaningful for Minimax to believe that the mare was demonstrating some empathy towards him.

Whilst it is not the case that all EAT/L centres follow a natural horsemanship approach, the philosophy and practice of natural horsemanship adopted by The Yard followed principles of respect and unconditional positive regard in terms of its horse management. Deana explains that for her this is evidenced by the fact that

> there's a respect for their...the horses' spirits which in my experience with kind of a lot of traditional stables or schools is that they're objects to be used by whoever, whereas here they're very much individuals, um, and respected for who they are.
>
> (Deana, therapist)

Perhaps by the practitioners demonstrating principles of 'unconditional positive regard' towards the horses, together with the fact that horses, as with other animals, are unconditional and congruent in their behaviour, an environment is created where the young people are able to experience this principle. Chardonnens suggests that

> a horse as a co-therapist presents, by the virtue of its characteristics and reactions, important similarities to the therapeutic conditions described as necessary and sufficient for the therapist to facilitate the change process.
>
> (Chardonnens, 2009: 327)

It may be that by building an authentic relationship with a horse, within an environment of unconditional positive regard, Kelly was finally able to build a relationship with her therapist, in the same way that Chardonnens suggests that 'the animal helps the therapist build an authentic relationship with the child' (2009: 328). Other studies claim to show the 'personal interaction with the therapist as the single most important factor in their treatment' (Sloane *et al.*, 1975: 225). These authors report that patients who felt high levels of warmth, empathy and genuineness in their therapists had higher success rates from therapy. It could be that the young person perceives and feels the horse as providing some of these characteristics, which then enables the growth of a relationship with the EAT/L therapist.

130 *Equine-Assisted Therapy and Learning with At-Risk Young People*

The psychotherapist Brooks states simply that one of the reasons why animals are more appealing to young people who have suffered abuse at the hands of adults is because they do not have any ulterior motives (Brooks, 2006). Deana suggests that the horse provides these conditions and characteristics because it

> doesn't read the referral form, doesn't have any preconceptions of this child who's coming, and have any idea of attachment disorder or therapeutic intervention or anything like that. Any label or ADHD or whatever. They meet the child in the present moment.
>
> (Deana, therapist)

The participant Lucy also voiced the opinion of how important it was for her that horses didn't label her. One day at The Yard, she described a difficult meeting she had recently attended regarding her disability living allowance (DLA) and a student work-experience placement at a local veterinary practice she wished to undertake as part of the college course in animal care she was enrolled on. She was clearly agitated and upset about the meeting when she arrived at The Yard, saying how confused she felt about being labelled disabled as that is not how she saw herself, and how 'the forms have been made by people who don't have a single clue!' However, once out riding on the horses in the countryside, she seemed to visibly relax and calm down. When asked what had facilitated this change, she replied:

> the horses don't judge you, they don't care what that man [the DLA officer] said earlier, they just take you for who you are.
>
> (Lucy, participant)

A further example of how the horses seemed to facilitate the therapeutic opening-up process is demonstrated by an extract from a TH session with Minimax. The Yard had been informed by his foster carer and social worker that he would often get very angry and upset about being separated from his younger brother, who was in the process of being adopted, and this could spiral into aggressive and violent behaviour. A hope for Minimax attending TH was that the calm and therapeutic environment would help him to become better able to express himself in more positive ways. After Minimax had been coming to The Yard for a while, he became more relaxed and confident around us, as described previously. The following section from fieldnotes shows how he then started to open up and talk about some of the things on his mind that were clearly bothering him, and how he turned to Ruby for comfort when this became too much.

Finding comfort: Minimax and Ruby

Once he was grooming Ruby we started chatting about Christmas and Minimax started talking very animatedly, which was quite a change as when he had started TH sessions he had initially been rather quiet and withdrawn in terms of talking about anything personal. Minimax told us about what he had been given for Christmas by his older brother and went on to talk about his brothers in more depth than previously. He said he was sad when he had seen his older brother before Christmas and that they were both thinking about their younger brother [with whom they had only letterbox contact now]. Minimax got quite upset and agitated at one stage, angrily spitting 'they [social workers] just don't understand, only Angie [his foster carer] understands what it's like.' We carried on quietly grooming the horses and giving Minimax the space to express his sadness about his brothers. During this time Ruby continued to stand calmly, not reacting to Minimax's agitation but turning her head to him from time to time to gently nuzzle him. Once he had calmed down, Minimax put down the brush he was holding in his hand and put his arms around Ruby's neck, hiding his head in her long mane and hugging her. He stood quietly like this for some time with just the rhythmical sound of the horses munching on their hay in the background.

(fieldnotes)

It appeared in this encounter that Minimax found solace and comfort from Ruby in some way and was able through this relationship to turn his anger about the powerless situation he was in to sadness at missing his brother. Ruby seemed to offer the unconditional, safe, physical body that he needed, and provide him with elements which he was unable to express or receive from people, at this stage of his life at least. The taken-for-granted nurture and comfort, both physical and psychological, which is available to the majority of young people from their families, immediate or extended, is not the case for young people in care who have been removed from their families. With the additional difficulties around touch within foster care regulations, due to the possibility of accusation of abuse (Pemberton, 2010), and the 'corporate parent' style of fostering, with relentless changes of social worker, carers and other people in their lives, the young person's need for actual physical touch and affection is often overlooked and lost. As was explored in Chapter 4 on the theme of attachment and further dimensions of tactile experiences with horses, it is suggested that young people who have been undernurtured can perhaps experience being 'held' by touching and hugging large animals (Brooks, 2006).

Body therapists, such as bioenergetic theorists, believe that certain childhood traumas and conflicts are 'embodied' in the patient – that is, by not being allowed to cry – and hence bottling emotions up and tension in the body can be a consequence (Kurtz, 1990). This can result in certain muscles becoming chronically tense, leading to tension/pain/distortion in other parts of the body. It is suggested that relieving these tensions through massage/body exercises/chiropody, and perhaps through activities with horses, can result in a discharge of these pent-up feelings and emotions. The re-emergence of hidden early traumatic memories can then be explored and resolved through psychotherapy (Brooks, 2006; Kurtz, 1990; Scott, 1980). Whilst these authors are talking about adults in the main, and the therapists at The Yard acknowledged that the young people were often too young, or their issues too 'raw', to be explored in any depth, we would discover that Minimax often complained of various aches and pains, especially backache, as did many of the young people who attended The Yard. In the instance given above of Minimax and Ruby, it may be that by grooming the mare and having this tactile experience with her, he was then able to relax, and this enabled the pent-up feelings about his younger brother to emerge. The area of tactile stimulation may also be additionally relevant in the case of Minimax as it was suggested that he was on the autistic spectrum. According to Scott (1980), this tactile element is especially important in the relationship between therapist and autistic clients where there are additional difficulties in attachment. Scott goes on to discuss the taboo of touch between therapist and patient and of how animals can therefore be valuable in this context, providing a safe arena to explore and experience touch and tactile elements previously missing or having held negative connotations in the patient's life (Scott, 1980).

Calm and safe: Horses and a safe space

A further theme which featured strongly in the study was of the horses appearing to be calming for many of the young people in various ways. This calming effect seemed to overlap with other themes, such as the importance of the natural environment, the practice of natural horsemanship, and the therapeutic philosophy and style of The Yard. The study found how feeling calm seemed to contribute to the therapeutic relationship between the young person and the horse, and the young person and other people.

Being calm appeared to be linked to feeling safe, which is considered a fundamental element of the therapeutic relationship (Frank & Frank, 1991; Friesen, 2010; Kurtz, 1990; Sedgewick, 2001). Fine, referring to AAT in general, adds that 'it appears that the presence of the animal allows the client a sense of comfort, which then promotes rapport in the therapeutic relationship' (2000:1 82). Esbjourn speaks of the 'calming effect' that she claimed horses appeared to have on participants in her study of EAP (Esbjourn, 2006). Whilst it is not possible to separate all of the different elements that are present within the relationship between young person, horse, practitioner and environment, the calming effect of a combination of factors did appear to be linked to enabling some of the young people to relax and open up, a precondition for successful therapeutic interaction. Although most of the data concerning the theme of 'being calm' was observational in nature, in a questionnaire they completed for the research, both Minimax and Cinderella commented on how the horses made them feel calmer. More often, however, the young people would refer to the horse's need to be calm, Kelly saying how 'they like it best when you are calm around them'. Being calm seemed to be linked to trust for Emma as she went on to describe how it was important 'to be quiet, to be calm. You've got to trust the horses as well as them trusting you, don't you?'

During one TH session with Cinderella, a three-year-old filly had just arrived at The Yard and was still a little anxious about having left her previous home and horses. When asked how she felt Sherry was feeling and the best way to respond to her anxiety, Cinderella replied: 'Well, she's probably scared and missing her mum'. She added that it was important to be 'calm and kind to her'. Freya also used the words 'kind' and 'calm' when she said that what she especially liked about the mare Ruby was that 'she looks sort of kind...and calm'. This suggested that Freya perceived Ruby as having a kind character, which was linked with the mare appearing to have a calm nature – the two were interlinked for her. Being seen to have a kind personality would seem to be a positive condition within the therapeutic alliance and perhaps a precondition to feeling safe and being able to trust. Developing a relationship with Ruby whom she perceived as kind and calm appeared to enable Freya

134   *Equine-Assisted Therapy and Learning with At-Risk Young People*

to start to develop a relationship with her human therapist, Deana. Yorke, Adams and Coady (2008) in their study with trauma victims participating in EAP speak of how 'Restoration of the trauma victim's capacity for recovery hinges on provision of safety and development of trust' (2008: 17).

On another occasion, being with the mare Ruby appeared to allow Kelly to feel safe enough to open up to Deana and express her emotions about a recent traumatic experience, something that she had previously not been able to do. This was something that Deana saw as a significant breakthrough, telling me that Kelly's normal instinct was to run away from her difficulties. The following is an extract from fieldnotes from that session.

Kelly assaulted: Opening up in a safe environment?

Kelly arrived and told us that she had been assaulted the previous day by a group of girls from her school. She moved quite slowly and stiffly but seemed pleased to be at The Yard and went over to say hello to Ruby, giving her a gentle stroke to which Ruby responded by turning her head to her and placing her muzzle in her hand. We gave Kelly the option of either taking Ruby for a short walk for exercise and to build on their relationship of starting to get to know and trust each other, or washing the horses' tails (as it was one of the first warm sunny days of the year). Kelly chose to take Ruby for a walk so we suggested she got the grooming kit from the shed in order to groom her first. She started to groom her but was very quiet, finally saying that she may not be able to clean out Ruby's hooves as it involved bending down and her back was stiff from the assault. We had touched on the assault a little before but now it felt as if Kelly wanted to talk about it some more. She talked about being jumped upon by eight or nine girls whilst others stood around filming it with their mobile phones. Deana and Kelly sat down at the bench and Kelly cried but the tears were silent. She said she hadn't wanted to come today and wanted to go back to The Elms. Deana rang The Elms for someone to come back to collect her earlier than the normal session end time. In the end, however, once she knew someone was on their way for her, she seemed to relax and ended up having fun throwing sticks for Toby [the dog] in the field, being in no hurry to leave when the staff member from The Elms arrived to collect her. Instead she spent time showing him the horses and telling him a little about their different characters, etc.

(fieldnotes)

Psychotherapeutic Insights 135

In her session case notes, the therpist Deana writes: 'It is possible that being in this environment allowed Kelly to relax her guard a little more than felt comfortable for her and her instinct was to run'.

(Deana therapy session notes)

In our post-session briefing after Kelly had gone, Deana explained how she saw this session as positive in that Kelly was able to start to open up in this environment. Even though this initially caused her to want to leave, when she realised that her escape route was not blocked she was then able to relax again and end on a positive note, not being in a hurry to end the session. Deana stated that she very firmly believed that Kelly's beginning to develop a trusting relationship with the mare Ruby was fundamental to this breakthrough in their therapeutic relationship, writing: 'there is certainly a relationship forming between the two of them'. She believed that, in turn, this enabled Kelly to start to feel safe enough with Deana to be able to express her emotions to her, which she normally kept very closed off. This is what Fine (2000) refers to when she talks about how animals can help clients to relax and how 'a therapist who conducts therapy with an animal present may appear less threatening and, consequently, the client may be more willing to reveal him – or herself' (2000: 183).

Many of the young people talked of how they believed it was important for the horses to feel safe. One way to help the horse to feel safe is by acknowledging their need to be able to move, and allow them space and a perceived escape route. This is perhaps similar to how this enabled Kelly to feel safer above. Often it seemed that the participants explored feelings related to safety through the horses, asking how they felt about certain things, such as leaving their previous homes, other horses, and how they behaved towards each other. In a semistructured interview, Emma was asked which horse she thought was the lead horse in the small herd. Emma replied that she thought it was Ruby:

EMMA: Because she doesn't put her ears flat back at them, she doesn't bare her teeth at them, she doesn't turn her bum on them. She just walks off from the argument sort of thing.

AUTHOR: Yes, she does, doesn't she?

EMMA: Jason, if you've noticed, if you let him, he just sort of barges his way through.

AUTHOR: *(laughs)* Yes, yes that's Jason . . .

EMMA: Because they know she's [Ruby] not going to be really aggressive and they know that she's not going to kick or bite or something.

136 *Equine-Assisted Therapy and Learning with At-Risk Young People*

> AUTHOR: Hm, hm. How do you think she makes the other horses feel by being like that?
> EMMA: Safe.

Perhaps through exploring the horse's behaviours and how she saw them relating to each other, Emma was in fact exploring aspects of her own psyche and feelings. Klontz *et al.* talk of how horses 'serve as catalysts and metaphors to allow clinical issues to surface' (2007: 258). The powerful use of identification, analogy and metaphor will be explored more fully later in this chapter.

The adult's views: Being calm, safety and trust

Some authors make the connection between aspects of EAT/L and Winnicott's 'holding environment' (Brooks, 2006; Esbjourn, 2006; Lentini & Knox, 2009; Vidrine, Owen-Smith & Faulkner, 2002). Esbjorn (2006) in her research on therapists' views of EAP found that many of them believed that the horses could provide aspects of the 'good enough mothering' that Winnicott had proposed. In the following extract from an interview with the therapist Deana, she also explains that in her view Winnecottian theories of object presentation and the holding environment are relevant to the theme of safety and trust within the therapeutic relationship in EAT/L and TH.

> Well, I think the thing that comes up to me is how...how resonant working with the horses is to...child development and object relations theory in particular, and I remember being out with you one day when we took Ellie [horse] when she was very young, and you were introducing her to objects in the environment like green wheelie bins and things like that um and we got talking about it and you talked more about your experiences and more about Lucy's article and reading that article, for me it's just pure kind of Donald Winnicott object relations theory of object presentation that the horse looks to you as a child would look to the mother to ascertain whether whatever it is in the environment is safe. And it was an amazing sort of mirror, really, that relationship between you and Ellie, and a mother and an infant...but you know, it's about trust, building trust, and her looking to you and asking, 'is this going to eat me?' or 'is this safe?' and, you know, you demonstrating to her that this is a safe world...And I think for a lot of these children that has not been their experience. And so there's something that goes on,

Psychotherapeutic Insights 137

I think, within the horse handler, the therapist's relationship with the horse and with the child that facilitates that reworking of that old model that the world is not safe.

(Deana, therapist)

Whilst the young people themselves spoke about the themes concerning 'being calm' and 'safe' more in terms of the horses, many of the adults involved in the young people's care talked of how the participants would often be calmer after attending TH sessions. Peter, a mentor from the YOT who brought Wayne for sessions, described how closed Wayne would be on his way to The Yard in contrast with their journey back to his foster home, when he would be much more open and talkative. In a questionnaire that some of the adults completed about TH, Peter wrote: 'He [Wayne] was always noticeably more relaxed as we drove away than when we arrived'. A foster carer, David, on the same questionnaire wrote about another participant '[He] is always calm on his return...there is a marked difference in his return in contrast to the anxiety he sometimes demonstrates at the start of the day'. Carl, the manager of The Elms, the residential children's home where Kelly lived, wrote that Kelly would often have 'a positive relaxed attitude' after a session. It is suggested that the very nature of therapy can create anxiety for some young people (Bowers & Macdonald, 2001; Brooks, 2006; Ewing *et al.*, 2007) and that introducing contact with animals within the therapeutic environment can help to provide a calming effect, especially during the initial sessions (Ewing *et al.*, 2007; Fine, 2000; Friesen, 2010; Walsh, 2009). Laura, a foster carer for a number of young people who attended The Yard, believed that there was a link between their being relaxed with the horses and this enabled them to open up. In the questionnaire she wrote:

The children I have looked after all enjoy the time spent with the horses. Because they are more relaxed they may be more inclined to talk about any problems or worries they may have.

(Laura, foster carer)

One reason why horses may help to facilitate a calm environment which, in turn, could contribute to the young people relaxing is that horses inherently want a calm and peaceful existence as they need to conserve energy for when it may be required in order to escape potential predators. As discussed previously, the horse will always look for a calm and intelligent leader that they can trust and feel safe with. Therefore

138   *Equine-Assisted Therapy and Learning with At-Risk Young People*

in order to have a successful relationship with a horse, it is necessary to behave in a calm and trustworthy manner (Blake, 1975; MacSwiney, 1987; Rashid, 2004; Rees, 1984).

Metaphor, analogy, projection and identification

The power and use of analogy and metaphor to change ideas and affect behaviour is well established, in both psychotherapy and other forms of communication. Gordon states that 'as long as recorded history has been around, and in myths that date back into the farthest and dimmest memories of human existence, metaphor has been used as a mechanism of teaching and changing ideas' (1978: xi). Within psychotherapy, Freud (1920) used sexual symbolism to explore unconscious fantasies and dreams, whilst Jung developed the terms 'animus' and 'anima' to describe male and female aspects of the psyche, and often employed animals as metaphors and symbols of unconscious feelings and meanings (Jung, 1978). In their work with torture victims at a therapeutic gardening project, the therapists use the power of metaphor and analogy to help clients come to terms with their past (Linden & Grute, 2002), and there are many parallels with the practice of The Yard. The authors of the therapeutic gardening project explain that

> Because the psychological work for victims of torture is so difficult to express in words, using nature as a frame of reference and a source of analogy and metaphor is very helpful for those with language difficulties.
>
> (Linden & Grute, 2002: 42)

Clients attending the Natural Growth Centre (NGC) possess many similar emotional and psychological scars to those of the young people referred to The Yard. Linden and Grute (2002) write about how clients at the NGC have difficulties with language barriers in addition to their psychological problems of expression. Whilst the young people in the TH study did not have the same language barriers to overcome, they did often have trouble in expressing and processing how they felt, partly due to their age in addition to their experiences. Angie, Minimax's foster carer, explained how she felt that there were similarities between horses and children in care in terms of their ability to communicate, which could perhaps help some of these young people to identify with them on some level, stating: 'horses haven't got words either, neither have foster kids, they don't have the words for the feelings'.

Psychotherapeutic Insights 139

The past experiences of some of the participants of The Yard and the clients of the NGC, although in different contexts, both contained terror, loss and abuse of some kind which can result in similar psychological distress:

> The future is as difficult for refugees to contemplate as the past because they have had the bitter experience of having their futures ripped away from them. Any future plans they may once have had – plans concerning their studies, their working lives, their homes, their families – have been obliterated. This can leave them with a sense of futility as regards making new plans, since experience had taught them that at any moment these plans may be disrupted.
>
> (Linden & Grute, 2002: 41)

It was experienced time and time again that many of the young people who attended The Yard moved placements with little or no planning or consultation. This left them powerless, anxious and traumatised, and this is demonstrated in the following section from sessional notes of a TH session during which Kelly was informed with no notice that she would be leaving her residential placement to move to a foster placement. In addition to being very traumatising for her, this left no opportunity for a planned and structured ending to the TH sessions.

Kelly's last session

On our return to the yard (from a ride in the lanes) the manager of The Elms was waiting for us. He informed Kelly that he had just received a call from her social worker (SW) explaining that Kelly needed to take all her belongings with her from The Elms to her mother's the next day as she would be moving directly into her new foster care home from there. Kelly was clearly shocked by this news. She was adamant that she wouldn't be going and wanted to speak to her SW. The manager rang the SW from the yard and Kelly spoke to her. Kelly reminded the SW of the original agreement, i.e. that if she didn't like the foster placement she wouldn't be obliged to go there. She explained that there was a younger child in the placement whose physical safety she had concerns for as she had previously assaulted a younger child. Kelly cried openly and appeared very sad and disheartened. We were sad when she drove off for the last time. It wasn't a positive ending for the sessions.

(therapy sessional notes)

140 *Equine-Assisted Therapy and Learning with At-Risk Young People*

This extract provides an example of the additional anxiety and trauma that some of the young people who attended The Yard were subjected to within their often disjointed lives in the foster care system, on top of the other traumatic experiences of their childhoods.

In order to help the participants come to terms with some of the negative experiences from their pasts, sometimes embodying and expressing this as emotional/behavioural problems, The Yard hoped that the experiences provided by the horses would give the young people the opportunity to get away from a preoccupation with their pasts and to be able to be more fully, and healthily, in the present. Linden & Grute, writing about clients at the NGC, explain that

> Medical Foundation clients are overwhelmed by their past experiences and cannot find a way to be in the present reality. They cannot concentrate on anything in the present and therefore cannot function well... the past is ever present.
>
> (Linden & Grute, 2002: 39)

They go on to suggest that the environment and being in nature can help people suffering from trauma to be brought back into the present. This can be done by 'focusing them on where they are – drawing attention to the grass, the sky, the birds singing' (Linden & Grute, 2002: 9). In the case of The Yard, the powerful experience of being in the presence of a horse that lives instinctively and fully in the present demands that, in order to be effective and safe, the handler too must be in the present. Gestalt therapy concentrates on how mind and body are interconnected, on learning to listen to instinct, and of being in the present (Clarkson, 1999; Perls, Hefferline & Goodman, 1972). Being with horses naturally incorporates all of these elements and so has many parallels to gestalt therapeutic approaches. Once the young person is in the present, the horse can act as a powerful metaphor for how they may be feeling and can ultimately, hopefully, serve as an unconscious way of helping them to heal themselves. Therapists at the NGC use the analogy of growing plants in a similar way, suggesting that

> When potatoes are sown, clients understand that they have to be looked after for the future and this is analogous with the clients' need to look after themselves in order to be healthy and 'grow'.
>
> (Linden & Grute, 2002: 41)

Psychotherapeutic Insights   141

Horses offer the power of metaphor and analogy in similar ways. Looking after horses is a very time-consuming, physical and in-the-present task. This experience of caring for another animal can serve as both a conscious and an unconscious metaphor for the young person caring for themselves both emotionally and physically (Bowers & Macdonald, 2001; Ewing *et al.*, 2007; Vidrine, Owen-Smith & Faulkner, 2002). Many comparisons regarding how it is important to look after aspects of the horses which mirror their own needs can be drawn, such as giving the horses rugs to keep warm in winter, checking the fields to keep them safe and maintaining their correct diets. Certain horses may need additional or limited feeding depending on their individual weight issues, for example. Some young people thought that it was 'cruel' to give certain overweight horses less food, and Cinderella wanted to give Leo 'a bit extra because he is my favourite'. However, when explained that rationing feed in this instance was for the pony's own well-being and health, this can serve as a valuable analogy in learning self-restraint and responsibility in other areas of the young peoples' lives. Minimax's foster carer also commented on how learning about horses had helped him to modify his behaviour, telling me:

> yes, ... they must understand how the horse is feeling ... yeah, it feeds back to them that maybe people feel like that ... And also he learned that if the horses weren't allowed to do something for a very good reason then he shouldn't ...
>
> (Angie, foster carer)

An example of how Minimax experienced this for himself is provided by the extract below from a session where he learnt why he needed to prevent the mare Ruby from doing something that she wanted (eating grass from the verge whilst walking in the lanes) for 'her own good'. He was then able to transfer this to an understanding of his own behaviour – an example of how the horse provided a metaphor which Minimax could understand.

'It's for your own good, Ruby'

Once back in the yard Minimax headed straight to Ruby and put his arms around her neck, giving her a hug. He stayed hugging her for some time whilst we carried on with yard tasks around him until he eventually joined us in grooming the horses, although he obviously preferred to hug Ruby than groom her! As we still had a little time left over we asked if Minimax would like to lead Ruby up the track with us as we needed to take Leo for a walk for some of his rehabilitation exercise. Minimax found this a little challenging as Ruby took the opportunity of Leo walking very slowly to stop and try and eat grass on the verges, dragging Minimax towards the tastiest patches and being reluctant to walk on when he asked her to. We asked Minimax what he thought he may need to do to help encourage her and he replied 'She's like me when I don't want to go to school, sometimes I want to stop and do stuff on the way and Angie gets annoyed with me!' We talked for a while about why it may not be good for Ruby to drag Minimax where she wanted and not listen to the person who was leading her, and what this could lead to if it was somewhere dangerous and the leader did not have control of her, Minimax correctly identifying that it could be dangerous if it happened on the road with cars around for example. With a little practice he soon changed his body language to being firmer and more assertive with her, identifying when she was starting to think about heading for the grass and preventing her before she did. He told her: 'It's for your own good Ruby'.

(fieldnotes)

Identifying with the horses

A useful and unique way in which horses can offer participants in EAT/L and TH the opportunity to explore internal issues is provided by their particular characteristics. Horses have very distinct personalities and positions in the herd, together with different past experiences, and it seemed that these often mirrored the young people's own personalities and experiences in some way. For example, the lead mare is looked up to, respected and followed by the other members of the herd, whilst other horses are 'bottom' of the herd when it comes to food/shade/water and so on when these are limited, especially in domesticated horses. Some horses are calm and gentle, some bold and curious and others more anxious and nervous, either due to their own personal characteristics and training and treatment in the past, or, most probably, a combination of the various factors. In addition, a number of the horses at The Yard were young and inexperienced and needed to be taught how to behave by people and the other horses. Likewise a number of horses were older and more experienced and had 'been round the block', as Lucy put it when talking about the older, lead mare Duchess. This combination of personalities and different pasts and behaviours provided a rich opportunity to

Psychotherapeutic Insights   143

use analogy and metaphor with participants in a non-confrontational way, and its uniqueness within EAT/L over other therapies is referred to by many authors (Karol, 2007; Klontz *et al.*, 2007; McCormick & McCormick, 1997; Trotter *et al.*, 2008; Vidrine, Owen-Smith & Faulkner, 2002). Questions about the horses to draw comparisons were sometimes asked if it was felt appropriate, but often young people would make connections and comparisons without any prompting, such as when Cinderella exclaimed: 'She [Duchess] is just like my *mother*, stubborn old cow!' This was on an occasion when Duchess kept walking away from her in the field, not allowing Cinderella to catch her. The incident led to a valuable TH session where Cinderella learnt that by changing her body language and approach, Duchess willingly allowed herself to be caught, as described in Chapter 6.

Some examples of questions sometimes employed at The Yard to help facilitate the process of bringing self-awareness to participants are given below:

- How do you think Jason/Duchess etc. is feeling today? How are they telling you? What could we do to help Jason/Duchess etc. be less irritable/happier? What do you think Jason/Duchess etc. needs to be happier/less irritable?
- Which horse do you think you are most like/is most like you? What is it about this horse which is similar to you? How do you think it feels to be this horse?
- Which horse would you like to be most like? What is it about them that makes you want to be like them? What would you have to do/be like to be like this horse?

The idea behind this style of questioning is to encourage the participant to find their own answers to any problems and issues that they may be experiencing, but in a way which is safer for them because the focus is on the horse and not themselves. It is important to note, however, that some of these questions would only be employed if the young person had initiated interest and curiosity about the horse themselves, in line with the non-directive, person-centred approach of The Yard, and together with a qualified and experienced therapist.

'I like him because he's naughty!'

Despite Emma supposedly coming to The Yard to look after and ride Timmy, the pony her adoptive mother had loaned to The Yard, Emma

144   Equine-Assisted Therapy and Learning with At-Risk Young People

soon developed a strong connection and relationship with the horse Jason. This coincided with her getting to know him and his particularly strong character over the course of a few weeks. Emma said that she especially liked Jason because 'he is grumpy' and would very much like it when he put his ears back in annoyance when we were changing his rug or doing something else which displeased him in some way. She would laugh and say: 'Look, he doesn't do that to me!' One day Emma was describing how her favourite horse at the agricultural college she was attending one day a week for work experience was 'the naughtiest horse there'. When asked why she thought she liked the 'naughtiest' horses best, she answered:

> well it's like apparently, but I don't believe this...umm, but anyway, apparently horses can read your mind from a mile away so I think they can tell that I'm bad...so it's like they know I'm on their side.
>
> (Emma, participant)

It seemed that Emma identified with the 'naughty' or what she perceived as the 'bad' side of Jason in some way. In the following extract from field notes, she described some more similarities she thought she had with Jason when this theme was revisited with her out on a hack.

Emma and Jason: Similar personalities?

Out on the ride Emma said she had thought of some more similarities between Jason and herself. They were that he loved attention and that he was very good at 'telling what they [people] are really like', which she told me she was very good at too, being able to tell if someone really was nice or just pretending to be. She asked me what Jason (who was now retired from riding) used to be like to ride and I replied that he was really lovely but hated schooling and doing dressage and that he got bored very quickly. Emma squealed with delight and laughed: 'That's exactly like me too, I get, like, totally bored in school, that's why I'm really naughty to the teachers and wind them up, just like Jason!' This obviously really pleased her and when we got back to the yard she immediately asked if she could catch him to put him into the pen for his dinner. As she went to catch him he tried to follow me with his haynet but quickly stopped when Emma asked him to and put his head down helpfully so that she could reach to put his headcollar on. Later on, whilst I was finishing putting things away in the shed and after Emma had finished all her tasks with Timmy, she went and got a brush and brushed Jason's mane until I was finished and had to ask her to stop. 'Okay, in a minute, just let me finish brushing his mane. Look how handsome you look now Jason,' she told him as he stood there willingly loose in the yard, allowing her to brush him even though she was using quite a hard hairbrush.

I noticed that they both seem to really be developing a relationship and wondered if it was fulfilling some need for Emma that she was giving Jason attention now that he was semi-retired (which means he doesn't get as much attention as he used to). It seems that there is something going on as Emma knows that she can never ride him properly and, as she has told me that her favourite thing about horses is riding them, then her affection towards Jason is obviously about something else. Is it the fact that she sees him as being so like her she is so obviously pleased about?.

(fieldnotes)

It may be that by Emma initially being attracted to Jason due to perceiving their personalities as similar and as being on the same 'side' in some way, she was then able to engage in a relationship with him that was based on caring for him. As Emma had previously stated that her primary motivation for being with horses was being able to ride them rather than the other tasks involved in horse management, which she often shied away from, this relationship, based on her identifying with Jason, facilitated her being able to obtain some possible benefits from what has been termed 'acts of helpfulness' towards another being.

Rejection and projection?

The participant Minimax initially had the opposite reaction to Jason, the two of them seeming to take a dislike to each other on their first encounter. Both his foster carer and his social worker told us that Minimax found reading social cues and body language difficult. This might have been related to his being labelled as having traits of the autism spectrum. He also had a diagnosis of ADHD and was on medication. For whatever the reasons, it was apparent that at his first session of meeting the horses, after he had got over his initial apprehension of being in a new place with new people, and perhaps in part down to nervousness, Minimax was rather overexuberant in his manner around the horses. He moved rather too quickly around them and tried to pat

146   *Equine-Assisted Therapy and Learning with At-Risk Young People*

them too hard on their faces, which, whilst some horses will tolerate it, they generally don't like. In this case, Ruby and Duchess did put up with Minimax's behaviour, but Jason, having a low tolerance threshold and a very strong character, demanding respect from his handlers, very quickly made it known that he didn't approve of Minimax behaving this way towards him. He displayed his displeasure by putting his ears flat back and shaking his head at Minimax in a dominant and threatening manner. Minimax immediately understood this body language and backed away, saying: 'Ooh, who's a grumpy boy then?' He seemed a little cautious of Jason after this and stayed away from him for the rest of the session, which was reciprocated by Jason. My fieldnotes after this session explore what was happening:

> I started wondering here about parallels between Jason and Minimax. Jason very clearly doesn't tolerate people who are too 'in your face' or not 'in their body' (by which I mean people who are not very 'grounded' or self-aware about personal space boundaries). Jason demands that people are very respectful of him, confident but not dominant, but he can also be very dominant, pushy, disrespectful of space boundaries, and 'in your face' when he wants something! The two of them quite clearly seemed to 'push each other's buttons', to use psychotherapy speak. Is Minimax now avoiding Jason because he doesn't like the part of Jason which is like himself?
>
> (fieldnotes)

Whilst there is clearly a human-centric interpretation in the effort to understand what may have been occurring between Jason and Minimax, certainly it appeared that Minimax did not like the part of Jason's behaviour which was in some way similar to his own as he avoided him after this encounter. Regardless of what is 'really' occurring it is through this sort of interaction with horses that learning and change can happen according to many practitioners and authors (Brooks, 2006; Ewing *et al.*, 2007; Lentini & Knox, 2009; Rector, 2005; Rothe *et al.*, 2005).

Interestingly, after Minimax had been attending The Yard for a few months, he began to show curiosity towards Jason again and wanted to start working with and riding him. This corresponded with Minimax's skills and self-confidence growing, alongside his ability to monitor and control his own behaviour and body language with the horses. During a short informal field interview where Minimax was asked some general questions around what he thought about the horses and their different personalities, he replied that he liked Ruby because she was 'quiet and understood' him, but also that 'I like Jason 'coz he's got attitude!' Asked if there were any other reasons why he liked Jason, Minimax went on to talk about how Jason behaved with some of the other horses, saying: 'He's like me at school. Like sometimes when I get angry with some of the other boys when they say things and stuff'. When Minimax was asked what he thought he had learnt from being with the horses, he replied: 'Well, you've got to think before you do things, it's like otherwise they might not understand...And they have to learn to trust you too'.

It seemed that the theme of trust had come into the relationship between the horse and young person again. Perhaps by Minimax having built up a relationship and trusting and feeling safe with Ruby he was then able to have the confidence to return to trying to build a relationship with Jason, the horse that had initially caused such a strong reaction for both of them. In their study of an equine-facilitated group psychotherapy vaulting programme with 'at-risk' young people, Vidrine, Owen-Smith and Faulkner make a similar suggestion, stating that

> this sense of being valued by a nonjudgmental other appeared to be related to the experience of feeling safe with the horse. This in turn appeared to be linked to constructive risk-taking in a cyclical relationship. (2002: 601)

Shared histories?

The participant Wayne appeared to be especially drawn to new horses arriving at The Yard. The Yard had an unusually busy time during the period of Wayne attending TH, with three or four new and young horses joining and passing through the centre for various reasons. One horse, Missy, was on loan from a horse-rescue charity but unfortunately, due to bad treatment in the past, was unsuitable for TH as she would bite and was unable to fit in with the other horses. This had the implication of rendering The Yard's insurance invalid. It was a difficult decision to send her back to the charity because obviously The Yard felt a responsibility towards her, understanding how hard it is for horses to constantly move home and not be able to secure bonds with other horses, which are so crucially important for them. There were also many parallels with the lives of the young people in care who attended The Yard, and this made it especially difficult because the practitioners did not want to be seen to be 'giving up' on a horse, as many of the young people may perhaps have felt that this had happened to them. The Yard wanted to model a

148　*Equine-Assisted Therapy and Learning with At-Risk Young People*

different approach, one of giving the horses a safe, secure home and a fresh chance, which they had been able to do with the mare Ruby, for example. Fieldnotes from around this difficult time describe how there appeared to be many similarities between Wayne and Missy, and how it seemed that he wanted to explore some of these through the mare.

Wayne and Missy: Exploring pasts

Beautiful sunny but crisp and cold spring day at The Yard again today. Missy still very grumpy and anxious on the yard so I put her in the wood pen separate from the others so she could eat her haynet in peace. Wayne and Peter (his YOT mentor) arrive early and when we ask Peter if he will be staying or walking his dog this time he replies that he will be staying but would like to walk his dog across the fields first if that is okay. As Wayne seems to find it difficult being around lots of adults we agree that this is a good idea. With Peter gone we try and engage Wayne in conversation about the things he did on his last session and what he would like to do this time etc. Wayne again remains rather withdrawn and uncommunicative but does agree to come and meet the horses. He is immediately especially interested in Missy being separate from the others in the wood pen and wants to stay and talk about her and why it is that she is apart from the other horses. We explain to him that it is for Missy's own safety and well-being as she is being bullied by the other horses and not being allowed to eat her hay in peace. He asks whether she is staying with us or going back to the charity (as he remembers us talking about this possibility on his last session) and we talk about the fact that she is going back so they can hopefully find a home which will suit her better and where she will be happier. After we have left Missy with her hay, Wayne agrees to come and groom Ruby and take her up to the round pen for a leading lesson in order to start getting to know her and learn some horse handling skills. I am aware that Wayne has suffered a long history of neglect and has recently moved foster placement and wonder if this is why he (as was another young person earlier in the week) is so interested in Missy, because he is relating to her experience. I wonder if by exploring her situation it will help Wayne to process his own experiences in some way.

(fieldnotes)

Our experience with Wayne was that he clearly found it difficult to communicate with adults but through relating to Missy, whose experiences seemed to mirror certain aspects of his life, he was able to start opening up and exploring some of these issues through the mare. Writing about the use of metaphor in drama therapy, Jones (1996) explains that some drama therapists suggest that the unconscious connections made by the client through metaphor are more valid than bringing them to consciousness, which would detract from their power by rationalising them.

In line with the non-directive approach of The Yard, together with the fact that many of the young people who attended were clearly not ready to process some of their traumatic experiences, which were still often very raw for them, it was left for the young people to make any direct connections between themselves and the horses without intervention from the practitioners. An exception to this was when questions about the horses' personalities and how they were feeling and so on (as previously listed) would sometimes be introduced if it was felt appropriate in order to bring greater awareness about the young person's behaviour towards the horses. Indeed, directive work in a psychotherapeutic sense would be unethical unless it was within a psychotherapeutic relationship. Being a social worker and not a psychotherapist, I was careful to remain aware of possible disclosures and where young people may be demonstrating that they were ready to start exploring difficult issues, and to discuss this in personal supervision and in the debriefing meeting after a TH session, in order to make a decision as to the best course of action for the young person. It often felt as though there was an unspoken understanding that sometimes the young people were talking about aspects of themselves and their lives through the safety of the horse. On a number of occasions when it was felt that the young person was trying to explore issues which were outside the remit of a TH session, a recommendation to The Yard's play therapist and psychotherapist Deana was made. Issues of transference and the unique role of the horse may be relevant at this juncture. Transference and projection are where the client projects some of their own feelings and emotions onto another person, the therapist in the case of the psychodynamic relationship (Rycroft, 1972). The therapist then has to hold and understand which are their own and which are the client's feelings, and this can be a difficult and potentially dangerous process (Frank & Frank, 1991; McLoughlin, 1995). The horse may be especially useful in this respect as it can act as a 'pure' mirror, without any of the added complications of transference and counter-transference between the client and the human therapist. Trotter *et al.* simply claim that 'the horse provides the vehicle for the projection of a participant's unconscious worries and fears' (Trotter *et al.*, 2008: 267). Furthermore, by the horse 'modelling' positive characteristics, such as resilience, tolerance and patience,

150   *Equine-Assisted Therapy and Learning with At-Risk Young People*

the young person may be able to take some of these on themselves, in the same way as Frank and Frank (1991) suggest this is possible within the evocative, or more non-directive, facilitating therapies, on which the practice and philosophy of The Yard were based.

Another powerful example of how the horses 'modelled' certain characteristics and seemed to provide further opportunity for Wayne to explore and perhaps process aspects of his past is provided in the following extract. Wayne had recently been taken into foster care and had already experienced a change of placement. In addition, he had difficulties at school and with his behaviour, resulting in his becoming involved with the YOT. Wayne had attended The Yard for about two months by the time of the following example and had become more confident around the horses. In fact he was sometimes overconfident, wanting to try activities which were rather above his stage of ability, but this seemed to correspond with his becoming a little more communicative with the practitioners. Although he would still remain rather quiet when he first arrived, he would begin to open up once he was actually engaging with the horses.

Wayne: Identifying with Billy

Despite our concerns that he may not attend, as he had missed his session the previous week, Wayne did arrive and had a very powerful session today. He was immediately interested to see the new young 2yr old colt Billy [who had arrived from Spain the previous week] and wanted to go straight into the field to see him. Normally we would not practice having young people in close contact with young or nervous horses but on this occasion as we now knew Wayne, and because it appeared to be important for him, I made an exception. This was only after I had explained to Wayne that Billy was still untrained and nervous of people and therefore we needed to be especially aware of our body language in order not to frighten him, and for him to feel safe with us. Wayne stood quietly in the field asking a lot of questions and commented about how Billy must been through a lot of changes since leaving Spain 'and leaving his mum?' He asked if Billy's mother would be coming over from Spain to see him. I explained that horses of Billy's age who lived as a wild herd would naturally be independent by now as their dam would encourage them to go out into the world by themselves and form their own group of other young horses. Wayne likened this to having his own crowd of friends who he goes out with in town. Billy was grazing next to Jason and Wayne pointed out that maybe Jason was like a big brother to Billy and helping him to settle in. We agreed that this was a very good analogy. As we were talking Billy slowly grazed his way over to us finally ending up standing right next to Wayne. Wayne seemed pleased about this and remained very quiet and aware of his body language in order not to startle the young colt, who finally allowed Wayne to very gently stroke his neck. We stood there quietly not wanting to break this special moment, but also aware that Billy was very young and virtually wild and unhandled – which could have caused him to startle and possibly bite or kick out in fear. However, both Wayne and the colt remained calm and finally Billy wandered off back to graze next to Jason. We commented on how well Wayne had handled the colt through his very mindful and aware body language and Wayne seemed pleased about this, although it is very hard to read his mind as he gives so little away.

(fieldnotes)

Paul Shepard talks about how he believes that animals play a key role in the transition from childhood to adulthood. Perhaps an aspect of this was occurring between Wayne in the fieldnotes above with his interest in exploring the young horse Billy's transitions in his life. Shepard writes:

animals have a critical role in the shaping of personal identity and social consciousness. Among the first inhabitants of the mind's eye, they are basic to the development of speech and thought. Later they play a key role in the passage to adulthood.

(Shepard, 1996: 3)

Melson too claims that animals play a vital part in children's growth and learning about themselves and others. She cites numerous studies that show how children's first narratives are 'peppered with animals of all kinds' and of how they help children to express 'anger, aggression, excitement, fear, sadness, happiness, being "good" and being "bad"' (Melson, 2001: 133). This aspect of being 'bad' was something that Emma clearly expressed in her identifying with the horse Jason as being 'naughty' and 'bad'. Melson gives the example of the child psychoanalysts Bellack and Bellack who devised tests which incorporated inkblots, and of how over half of the children whom they analysed in this way saw animals in the inkblots. They concluded that this childhood propensity towards animals held a key to gaining insight into the child's inner world, relationships and conflicts in psychoanalysis

152 *Equine-Assisted Therapy and Learning with At-Risk Young People*

and that 'animals may be preferred identification figures from three years to possibly ten' (cited in Melson, 2001: 137). Whilst the majority of the young people who attended The Yard and participated in TH were in their teens, it was generally accepted by the professionals and other adults who referred them that, due to their abusive and traumatic pasts, they were often emotionally developmentally delayed, acting out behaviours of development far younger than their actual years.

The psychoanalytic terms 'identification' and 'introjections' may be useful in terms of attempting to understand the intricacies of what is happening within the therapeutic space between the horse and the young person in TH sessions. Introjections are evaluations – negative in this context – about the self which are internalised and become part of how the individual defines and sees themselves (Mearns & Thorne, 2000). Introjections such as 'I am useless' and 'I am bad' were common with many of the participants who attended The Yard, Emma being an example cited earlier with her introjections that she was 'bad', which is why she felt that Jason understood her. In the session 'Kelly's last day' described earlier, Kelly told her social worker of how she was frightened of what she might do to a younger child in the foster placement that she had been told she was moving to as she had previously assaulted a younger child, clearly indicating to the social worker how she was scared of this 'bad' side that she felt she possessed, and over which she felt she had no control. It was perhaps interesting that the participants Cinderella and Minimax chose these characters as their pseudonyms. Mearns and Thorne (2000) talk of how configurations built around introjections can resemble 'fairy tales'. They give the very example of Cinderella 'who could be wonderful if only she could find her "prince"' and 'the boy who could be the "comic strip hero" if only he could do better than he can' (Mearns & Thorne, 2000: 109). Minimax took his name from a toy character and certainly had problems with his behaviour at school, which his carer said had contributed to his being academically behind other children of a similar age.

Conclusion

It appears that there were a multitude of factors occurring in the interactions between the horses and the young people, with it being very difficult to separate these out. Psychotherapeutic themes of identification, metaphor, introjections and Winnicott's safe 'holding' environment, alongside Rogerian principles of unconditional positive regard, are argued to be especially relevant in this chapter and examples are

Psychotherapeutic Insights   153

given of their application. The data in this chapter have drawn more heavily on the views of therapists and other adults involved in The Yard, as this is an area that they particularly represented. How the young people expressed themes related to psychotherapeutic processes was more often through the horses – for example, they often spoke of how the horses needed to be calm and safe, and how trust was important on both sides of the relationship.

Whatever the guiding processes were, it emerged that many of the participants were able to relax and open up in the presence of the horses. Whether it was a combination of being in the natural environment, the philosophy and style of the practitioners of The Yard, the natural horsemanship ethos, or merely just being with horses, it nevertheless appeared that the horses were the 'glue' or motivating factor that brought these young people back week after week, and the relationships that they developed with them that facilitated the start of the therapeutic process for many of them. Chapter 6 explores another area related to the therapeutic literature – that of the practice of mindfulness and its growing application in the therapy field, alongside its links to the natural environment and the expanding field of nature or ecotherapy.

6
Horses, Mindfulness and the Natural Environment

Introduction

As previously described, a major theme running throughout the research study, related to both the young people and the adults, was based around the words 'calm' and 'relaxed'. For the young people this seemed most strongly linked to the horses; for the adults, references to being in nature also featured quite strongly. The themes of being in nature and relaxing with the horses has links to the nature therapy literature (Berger & McLeod, 2006; Coleman, 2006; Peacock, Hind & Petty, 2007), and this is addressed in more depth later in this chapter, together with other suggested benefits from being in the natural environment with horses, such as the learning and physical benefits. First this chapter looks at how both being with horses and being in nature can perhaps bring people closer to experiencing what is described as mindfulness, in the modern Western concept of the practice. This is illustrated through examples from the data which show how the young people were able to adapt their behaviour, which sometimes appeared risky, and demonstrated an unawareness of their actions, in order to have effective relationships and experiences with the horses. To do this, the young people needed to become more aware of their body language and how their emotions and feelings affected the horses, whilst at the same time being relaxed and focused – similar principles to those described in some of the mindfulness literature.

Mindfulness, child psychotherapy and TH

Mindfulness within Western health models has evolved from ancient meditative traditions and has its roots in Buddhism. In its simplest definition it is the practice of relaxed concentration in the present moment, as opposed to ruminating on past events or being preoccupied with the future (Biegel *et al.*, 2009; Brown & Ryan, 2003; Moss, Waugh & Barnes, 2008). Mindfulness is suggested as being a method of observing 'what is happening right now, in our bodies, minds and the world around us' (Halliwell, 2010: 16).

Being safe and developing a successful partnership with horses requires being in the 'here and now', relaxed concentration and attention to body awareness. This is because the horse is constantly looking for direction from its handler/rider through its own means of communication, which is a keen awareness of body language and physiology due to millennia of being a prey animal. If the handler is distracted and not aware and present, the horse will quickly respond by taking control of the situation as it will feel unsafe. The sociologist Game, in describing riding, suggests that what is required is

> *relaxed* concentration, a very focused and meditative state. Maintaining connection and rhythm doesn't work through the exercise of will power, but requires a mindfully embodied way of being.
>
> (Game, 2001: 8)

Germer (2005) describes the opposite of mindfulness as 'mindlessness' and gives examples of this as including rushing through activities, carelessness, being unaware of tension and a preoccupation with the future or past. The young people referred to The Yard would frequently arrive in either a hyperactive state or, alternatively, rather distant and withdrawn and often unaware of how their behaviour affected the horses. Their social workers and carers informed us that many of the participants could become stuck on past events in their lives, either repeatedly going over these events or acting out behaviours. Emma's mother, Linda, explained that Emma had a Statement of SEN and a one-to-one teaching assistant due to her difficult behaviour at school. She went on: 'Yeah, you know, she'd go off or, you know, there has been quite a lot of time with [Emma[thinking of the past.' By being with the horses it was found that the young people would become more focused, attentive and aware of their body language, and, in turn, be able adapt this in order to have effective relationships with the horses. As many of the young people, like Emma, who attended The Yard found it difficult to engage in traditional educational and therapeutic approaches, it may be that EAT/L and TH practices, which naturally incorporate many techniques and practices in working with horses which parallel mindfulness practices, such

156 *Equine-Assisted Therapy and Learning with At-Risk Young People*

as MBSR, can offer those benefits but within a more informal approach and environment.

Horses and being calm

A core element of mindfulness practice is to obtain a calm and relaxed state of mind and body, whilst also being aware of thoughts and sensations without judgement. The physical health benefits of reducing stress levels are well documented in addition to the mental health benefits, so mindfulness-based exercises may be useful to help achieve this (Halliwell, 2010). Many of the young people openly expressed how the horses made them feel calmer. Feeling calm also seemed to be linked to feeling safe, with the participants often expressing this as the horses needing to feel calm and safe, and was another theme to emerge from the research. The theme of feeling safe is explored in more detail in relation to the therapeutic relationship between client and therapist in Chapter 5, but in this context it is linked to feelings of calmness. When asked how it made her feel being around the horses, Lucy replied:

> Um, when I'm angry, they make me feel a lot calmer because you have to be calm around them. You have to be calm and assertive around them, so, you kind of, you know, end up being like that. Well, I do anyway.
>
> (Lucy, participant)

She went on explain how learning about horse behaviour helped her to be able to work effectively with horses and how this also helped her to become less focused on negative past events at a difficult time in her childhood:

> and knowing exactly what you want to do and at all times because you're the leader. You're the herd leader. Yeah. And it kind of, it kind of helped me focus on something else rather than me.
>
> (Lucy, participant)

One of the questions in a questionnaire completed by some of the young people for the research study was 'How does being around the horses make you feel?' MiniMax wrote: 'The horses change your mood and they help to calm me down'. Cinderella put simply: 'They have a calming effect.'

The EAT/L practitioner and counsellor Sally referred to how she believed that being in the horses' environment is in itself relaxing and 'slows you down'. Going on to talk about EAT/L at the residential facility for young people with autism where she also worked, she said that she believes that the horses help to create a peaceful space where the young people can relax:

> And that stillness, I mean, if we weren't talking now we'd probably be sitting here quite quietly, wouldn't we? The horses have been pottering around quite quietly, and the whole thing just slows down. So, it seems to me that some of the youngsters really feel quite relaxed with the horses, like it's almost a relief in a way to be with a non-verbal being. They've had all the staff yacking through the day and other kids yacking, and now they're in a peaceful space with a non-yacking sensuous being who actually knows how they're feeling.
>
> (Sally, counsellor)

As described previously, mindfulness therapies such as MBSR and MBCT differ from traditional psychotherapeutic and analytic approaches that concentrate on looking at the client's 'problem'. They are, instead, more concerned with bringing attention to the body and encouraging relaxation, together with an emphasis on 'being mode' as opposed to 'doing mode' (Segal, Williams & Teasdale, 2002: 76). This approach is aligned to gestalt body therapies which have commonalities with some of the practices of The Yard where a keen awareness of body language, with exercises to encourage and facilitate this, was encouraged. Mindfulness practices understand that the suppression of negative emotions can cause stress (Brown & Ryan, 2003; Germer, 2005). Many of the young people who attended The Yard were unable to engage in traditional psychotherapy for a number of reasons. This could include their age, 'conditions' such as ADHD which made concentration difficult for them, and because their issues were perhaps too 'raw' and painful for them to explore them openly (Biegel *et al.*, 2009; Zylowska *et al.*, 2008). Instead it has been suggested that these young people may benefit more from exercises which MBSR techniques employ, such as stretches, mindful walking and yoga, in order to encourage calmness (Segal, Williams & Teasdale, 2002). An exercise which has similarities to these and included a 'body scan' was employed in the 'invisible riding' technique developed at The Yard. This exercise focuses on bringing awareness to each part of the body in turn in order to feel where tension may be being held. By relaxing these parts of the body increased sensitivity and connection with the horse at a deeper level can be achieved and the horse will be more comfortable and 'tuned' in to the rider as a result.

'Invisible' riding: Awareness and relaxation on the back of a horse

Because they are big, powerful and potentially dangerous, horses demand attentive, moment-to-moment concentration. Paradoxically, however, because they are also prey animals horses find attention concentrated too fully on themselves anxiety-provoking – this is how a predator would stalk them (Rees, 2009). Instead it is more important for the handler to be aware of their *own* physiology, body awareness, breathing patterns, and how they move around the horse, in order to convey a relaxed, confident, manner so that the horse will feel safe. A number of natural horsemanship and classical riding trainers incorporate yoga, Alexander technique and other practices involving body awareness and breathing exercises to assist in teaching more mindful body awareness (Bentley, 2001; Rolfe, 2007; Tottle, 1998). Drawing on links between attachment theory and mindfulness, Shaver *et al.* write about 'a stronger and wiser other who helps a client or seeker of emotional stability become less anxious, less avoidant, more secure, and more effectively mindful' (2007: 269). Perhaps a combination of these elements could be seen to be at play in the next section drawn from a TH session with Freya. In this session she found how achieving a state of mindfulness enabled her to connect with the mare Ruby in a way which was very meaningful to her, and which facilitated a profound change in her behaviour and manner, from potentially risky and harmful at the beginning of the session to calm and centred by the end.

Freya and Ruby: 'Invisible' riding

Freya arrived for her first TH session with the care worker from her residential home and initially seemed rather quiet and withdrawn, finding it difficult to engage or tell us what she would like to do with the horses. However, once her care worker had left Freya did began to join in with us putting on headcollars and grooming the horses but it soon became apparent that she had little awareness of personal safety around the horses. She put herself in quite dangerous positions and seemed oblivious to simple horse behaviour signals that other young people would usually instinctively possess. On one occasion

Horses, Mindfulness and the Natural Environment   159

when the horses were coming through a gateway and became a little agitated and bargy with each other, a clear signal to keep a safe distance until they had resolved their positioning, Freya continued to stand in the middle of the gateway, potentially allowing herself to be trampled. Because of this, and because her concentration levels on keeping on task with grooming and yard tasks were difficult for her we decided to take the unusual practice on a first session of suggesting Freya ride a horse today. By being on top of a horse, physically being in one place and connected to the horse, and also because she would need to be more aware of her body in order to find the balance and control needed to stay on, we hoped that Freya may be able to find some way of being able to concentrate and become more mindful of her behaviour.

When we put the suggestion to Freya to ride, she readily agreed. Freya's reaction again appeared to confirm to us the risk-taking behaviour that was stated on her referral form. Her therapist had relayed that Freya would often put herself at-risk in her life with her peers and with risky behaviours outside the residential home. Once up on the mare Ruby however, Freya's distracted behaviour appeared to wane a little and she became quieter in her body language and manner, seeming to suddenly realise that she was perhaps a little vulnerable on top of the horse. This appeared to enable her to listen to us and take instruction more readily, and we took the opportunity to introduce her to some 'invisible riding' techniques. In the round pen we initially led Freya around on Ruby, the mare seeming to understand that she needed to remain extra attentive and alert today, perhaps picking up on Freya's emotional and physical state. We started off with some simple stretching exercises in order to help enable Freya find her balance, tune in to the different parts of her body and gain some more confidence, as well as being fun. Next we introduced some simple 'body scan' exercises where Freya concentrated on each part of her body in turn, starting with relaxing her feet and moving up her body until she relaxed her shoulders and neck. In order to make this more fun and engaging I demonstrated these walking next to Freya, who copied the exercises riding on Ruby's back. Once Freya had found her balance and was more relaxed we suggested that she close her eyes and ride Ruby with her eyes shut in order to really tune in to Ruby's movement. This is not as easy as it may appear (try standing on one leg with your eyes shut!) but is a really useful exercise for refining balance and for following

160 *Equine-Assisted Therapy and Learning with At-Risk Young People*

the movement of the horse. Together with this we demonstrated to Freya how she could slow her breathing down, and breathe in and out in order to influence Ruby's pace, and learn how to bring her to halt and to walk on again just by the smallest body movement and breathing. This exercise takes a lot of sustained concentration and body awareness, together with real intention; it will not work unless you are completely committed and 'mindfully embodied'. After a few attempts we knew when Freya began to get a sense of this feeling as she gained more ability in co-ordinating her body language together with her breathing and concentration. In turn Freya's confidence in her newly found body awareness grew and Ruby responded accordingly, causing Freya to exclaim: 'Look, she slows down when I'm just thinking it now.' Later, as we finished the session, Freya stretched down from Ruby's back to hug her around her neck beaming: 'It's like she can read my mind.'

(sessional and fieldnotes)

The above session with Freya would seem to indicate that she achieved some of the elements which are described in the mindfulness literature. By becoming more present and 'in the here and now', together with incorporating some of the body exercises suggested as being important in order to achieve body awareness, Freya was able to leave her usual mode of being, which was to be unfocused, unaware and potentially harmful to herself, to find a new way of feeling and behaving. It may have links to what Segal, Williams and Teasdale suggest when they write about how a main aim of the body-scan exercise practiced in MBSR is to help participants to 'develop concentration, calmness, flexibility of attention, and mindfulness' (2002: 110). It appeared that the body scan and the other exercises undertaken together with the presence of the horse enabled Freya to achieve the levels of mindfulness, concentration and relaxation that she demonstrated during the 'invisible riding' session. Ruby would seem to have been the motivation that enabled Freya to participate in the process and may have acted as the 'stronger and wiser other' that Shaver *et al.* (2007) mention in their suggestion of the similarities between attachment and mindfulness. Freya certainly displayed affection towards Ruby at the end of the session when she stretched down to hug the mare, a bond which was not immediately apparent until she had experienced some connection and communication with her. There may also have been elements of Freya having experienced some aspects of the 'authentic functioning' that Heppner and Kernis (2007) and Brown, Ryan and Creswell (2007) propose, which is explored in more depth in the section with Cinderella and the mare Duchess later in this chapter. Freya was initially overconfident and displayed risky behaviour, which it is proposed by Heppner *et al.* (2008) can be a defence mechanism to hide underlying fear. Certainly, as Freya attended further TH sessions, she became able to show her lack of confidence around the horses and was able to overcome this in a more authentic way, which was safer and more meaningful for her. An example of this is provided in Chapter 3 in a round-pen session with Freya and the horse Hector.

Mindful horses

One of the reasons why horses may be useful in helping to help participants to achieve aspects of mindfulness is due to their particular psychological characteristics. As previously mentioned, horses are social prey animals, constantly on the look-out for potential danger despite millennia of domestication. They rely absolutely on the security of the herd and a trustworthy leader to ensure their survival (Budiansky, 1998; Kiley-Worthington, 1987; Rees, 1984). In the absence of another horse they will turn to a human to provide this, but, if the human is unable to, the horse will learn to rely on themselves to remain safe. This can often lead to unwanted and potentially dangerous behaviour as, under perceived threat, the horse may decide that it is safer to run away from the handler than stay with them, with potentially dangerous results for human, horse and, sometimes, whatever is in their way. Being herbivores, horses have to graze for the majority of the day so it would be inefficient for all of the horses of the herd to be constantly alert and on the look-out for danger. Therefore the stallion will generally be lookout together with different horses in the herd taking it in turns to be on alert. If danger is sensed the whole herd will react and flee until a respected, knowledgeable herd member will return to grazing demonstrating there is nothing to fear, for example. This may be demonstrated by a combination of raised heartbeat and breathing, minute change in muscle tension and perhaps the smell of sweat, as it is understood that horses have a highly developed sense of smell in addition to their other senses (Blake, 1975; Gehrke, 2006; Kiley-Worthington, 1987). It is therefore possible for human handlers to influence the horse's behaviour by learning to control their own body language, and to bring down a raised

162　*Equine-Assisted Therapy and Learning with At-Risk Young People*

heartbeat through breathing techniques and other mindful exercises. As the counsellor Sally described it,

> the thing that horses do really well is they're incredibly sensitive to body language and emotions. And although other animals are, it's not in the same way and they don't have that ability to mirror them. I mean, there's some confusions about this issue, I think, but the reality is that if you are afraid and uptight, the message to the horse is there is something to be afraid about, so the horse gets afraid and uptight too. And this works the other way too, so if you are calm and relaxed the horse is more likely to be as well.
>
> (Sally, counsellor)

Practising mindfulness, in terms of becoming 'mindfully embodied' and aware of one's own body language, is necessary to be able to work effectively with horses on the ground, as well as ride them. In the following example of a TH session with Ruby, Wayne was able to overcome a challenging and potentially frightening experience by practising mindfulness. This was in the context of being aware of his body language and breathing, and moderating them, in order to demonstrate to Ruby that there was nothing for her to fear.

Wayne and Ruby: Calm and confident

Wayne chose to work with the mare Ruby on this rather cold, windy, winter day. Despite the wind Wayne was keen to try the obstacle course in the arena as he had wanted to do this previously but had not had the opportunity. Before commencing the obstacle course we talked with Wayne about the reasons horses can get excitable and frightened in the wind and he correctly guessed that this is partly due to them not being able to hear as well in the wind. We went on to discuss the importance of acting 'calm and confident' in order to model a calming influence to the horse, who is always looking for a secure and safe leader to follow, whether this is horse or human. Following this discussion we taught Wayne the simple 'body scan' exercise of noticing his breathing pattern and of how to slow his breathing down in order to feel calmer and so mirror this to the horse. Wayne enjoyed placing his head against Ruby's warm body and feeling her slow breathing rate whilst she was at rest, then mirroring this himself so they were breathing in tandem. He also commented on how he thought 'she smells nice' whilst doing this exercise.

In this session in the arena Wayne had the opportunity to put this exercise of modelling 'calm and confident' and monitoring his breathing into practice when Toby (the dog) ran over the bank causing Ruby to jump sideways, snorting and almost resulting in Wayne dropping her leadrope and instinctively running away. However, once we reminded him of his body language with the simple prompt 'remember...calm and confident', Wayne was able to compose himself and remain with Ruby, standing quietly and calmly, then stroking her neck gently and telling her not to worry. As Ruby calmed down and turned to him for reassurance it would seem that this experience served a powerful message to Wayne that he was able to effectively control and manage his own behaviour and provide reassurance and comfort to another being, something that perhaps had not always been provided to him in his childhood. By being in the 'here and now' and aware of his body language and breathing Wayne could find that he could become 'calm and confident' with effective results.

(extended sessional notes)

In the examples of TH sessions with both Wayne and Freya where breathing and body-scan exercises were employed, there are clear links to the mindfulness practice of 'mindfulness walking'. Segal, Williams and Teasdale (2002) also describe this as 'meditation in motion', where participants are taught a technique of walking mindfully while keeping a 'soft gaze', and suggest that it can be a particularly useful exercise for people who feel agitated and find it difficult to 'settle'. These mindfulness practices draw on principles that for some participants it is easier to achieve mindfulness with 'a practice that involves physical movement than with one that does not' (Segal, Williams & Teasdale, 2002: 181). Coleman (2006) further believes that it is easier to come closer to mindfulness by being in the natural environment. Through a combination of being outside in the natural environment, in close contact with a 'stronger, wiser other' (Shaver *et al.*, 2007) and practising exercises with similarities to some of those described in the mindfulness literature, it would appear that participants such as Freya and Wayne, who sometimes found it difficult to concentrate and 'settle', were able to experience some of the benefits suggested as arising from achieving a sense of mindfulness. Morgan and Morgan (2005) write about how most of the time people are in a state of 'partial attention' and this can be more so for people who are stuck in patterns of negative thought due to past experiences. They suggest that experiencing

164　*Equine-Assisted Therapy and Learning with At-Risk Young People*

mindfulness and 'optimal presence' brings other benefits in terms of more positive relationships, and empathy and compassion towards others due to gaining an enhanced sense of well-being (Morgan & Morgan, 2005). Wayne certainly appeared to demonstrate some empathy towards Ruby's fear, and he was able to monitor his own fear and emotional response in order to remain with her, and eventually calm her down.

Self-esteem, authenticity, aggression and mindfulness

In this section it is demonstrated how the participant Cinderella learnt to adapt her behaviour, using some techniques aligned to some aspects of mindfulness practice in order to successfully catch and halter the mare Duchess. Heppner and Kernis (2007) and Brown, Ryan and Creswell (2007) write about how mindfulness can lead to 'authentic functioning'. Heppner and Kernis (2007) explain that authentic functioning is when individual gains increase self-knowledge and awareness, which can enable one to connect to their 'core-self'. Although a controversial concept, they claim that their research 'has shown that mindfulness and authenticity are interrelated' and that more positive relationships, lower stress levels and better mental health are a result of authentic functioning (Heppner & Kernis, 2007: 249). In addition, Chambers, Gullone and Allen (2009) make the link between mindfulness and emotion regulation, which could also be seen to have some relevance to the examples given in this chapter with a number of the participants.

As previously discussed, in order to have a successful partnership with a horse it is necessary to be self-aware and in control of your body language. Horses are incredibly sensitive and will respond unfavourably to aggressive or agitated behaviour, sensing even increased heart and breathing rates. Working with and learning about horses is inherently experiential; it is only possible to receive feedback through embodied practice with them – no amount of theoretical learning will achieve the same result. The following fieldnotes of a TH session with Cinderella showed how she demonstrated some of the behavioural traits of being highly sensitive to criticism described by Heppner *et al.* (2008) and Heppner and Kernis (2007). It is also shown how Cinderella gained some insight into how her negative thoughts and feelings affected her body language, which in turn affected the horse Duchess. She discovered how being with these feelings instead of repressing them, and relaxing, both in mind and body, then exploring how she could change her body language, resulted in her having a more positive experience with Duchess.

Cinderella and Duchess: 'She's just like my mother!'

Cinderella seemed very keen and enthusiastic to be back and came skipping into the yard today. She had bought her own leadrope, hoofpick and a crop from a riding shop local to her foster home. I didn't challenge this immediately but as she was waving the crop around rather erratically, I suggested that she put this in the tack room for now and later explained that we don't generally use or have crops at The Yard, pointing out one of the policies on the wall which stated 'No whips'. Cinderella initially looked rather deflated and then almost immediately became very defensive, saying: 'Well I won't bloody bother again then.' I quickly reassured her that we were very pleased that she was buying her own things for the horses but suggested that they might prefer carrots to whips next time! Cinderella seemed to perceive this as criticism, and obviously found it difficult, so I suggested that we put a special hook up for her to keep her own leadrope and hoofpick on especially for her to use with Leo when she came to The Yard. This seemed to help shift her black mood to an extent and she helped me put up the hook.

The five horses were out in the field so we took headcollars and walked out to catch them. We talked about which ones to catch first in order that the rest of the herd would follow and Cinderella correctly identified the two mares, Duchess and Ruby, 'It's them two isn't it... 'coz they're the ones the others look up to.' As Ruby was over at another part of the field, I asked Cinderella if she would like to catch Duchess and stood back to allow her to do this. However, Cinderella approached Duchess in a rather dominant, almost aggressive, manner, which caused the mare to walk purposefully away from her, refusing to be caught. Cinderella immediately got angry and frustrated, walking off and throwing the headcollar down then throwing herself down on the grass loudly exclaiming: 'Stubborn bloody cow. Don't be bloody caught then!' I sat down next to her and, to her surprise it seemed, praised her for her actions, telling her that sitting down and not chasing after Duchess was in fact a very good strategy and one of the tactics I may try with a horse who didn't want to be caught. I suggested that we sit in the field and relax for a while and just observe Duchess and the horses without necessarily trying to catch them, but at the same time bring some awareness as to how she felt Duchess may be feeling. After sitting quietly in silence for quite a long time with only the sound of birds and the wind in the trees in the background, and with Cinderella appearing to be ignoring me, she finally said 'She probably doesn't want to leave the others and I suppose she doesn't know me yet, but she's still a stubborn bloody cow . . . just like my mother she is!' I asked Cinderella what different approaches may help Duchess to want to be caught and she replied: 'Well, probably getting to know me a bit more first so she knows she can trust me.' We followed this with a short discussion about a horse's body language and whether Cinderella could see if there were any different approaches she could try to help Duchess learn to trust her. With this Cinderella agreed to try approaching Duchess together with me in a slower, more controlled, and less aggressive manner, and did then succeed in carefully putting the headcollar over her head. The other mare, Ruby, then followed us into the yard where Cinderella put the headcollar on her, also with no problem, her body language reflecting a much more gentle approach towards the horses, who responded accordingly.

(sessional and fieldnotes)

It would appear that Cinderella's feelings surrounding her mother (from whom she had been separated for some while, and with whom she had a very volatile and inconsistent relationship according to her social worker) were quite close to the surface, as demonstrated by her comparing Duchess to her mother in the heat of the moment after being rejected by the mare. Another claim that is perhaps related to Cinderella's experience with Duchess is that more mindful and authentic functioning can result in the participant becoming less aggressive as they begin to gain greater 'secure' self-esteem (Heppner & Kernis, 2007; Heppner, *et al.*, 2008). This claim is supported by Zylowska *et al.* (2008), who reported a reduction in aggressive behaviour in their study with adolescents assessed with conduct disorder who participated in a mindfulness programme. Where individuals possess fragile self-esteem due to their negative life experiences (such as many of the young people attending The Yard who had had to develop protective defences due to their traumatic pasts), they may be more likely to be highly vulnerable to challenge (Heppner *et al.*, 2008). The referral forms of some of the participants at The Yard spoke of their difficulties in taking instruction, of being ultrasensitive to criticism and of aggressive behaviour, traits which suggest fragile self-esteem. Cinderella's aggressive manner may have been linked to her possessing fragile self-esteem,

Horses, Mindfulness and the Natural Environment   167

or strong ego-involvement, and so she may have already been expecting rejection and failure with Duchess. Heppner *et al.* suggest that 'aggressive behaviour is one means by which people attempt to restore their damaged self images' (2008: 487). Cinderella's aggressive body language then resulted in her perceived failure in the task of catching and haltering Duchess and so her defence mechanism of further aggressive behaviour was triggered; 'social rejection or ostracism often leads to heightened aggressive behaviour' (Heppner *et al.*, 2008: 487). However, by allowing Cinderella to sit quietly and not judge herself, or the mare, negatively, but just relax and be with her feelings for a while, it appeared that Cinderella was able to explore and process some of her difficulties around trust and security, albeit possibly unconsciously. In turn this was reflected in her body language becoming gentler and softer with the mare and, as a consequence, being able to catch and halter her successfully, which in turn provided her with positive feedback to her actions. This would seem to have some connection with Segal, Williams and Teasdale's (2002) description of MBSR exercises which they suggest can help participants to gain greater awareness of how negative emotions can be expressed through the body. In addition, by being able to regulate her response there may be links to the 'mindful emotion regulation' that Chambers, Gullone and Allen propose when they suggest that it can 'allow the individual to more consciously choose those thoughts, emotions, and sensations they will identify with, rather than habitually reacting to them' (2009: 569). By engaging in embodied activities with the horses which demand authentic behaviour, as this is all that horses will positively respond to, it may be possible that participants achieve aspects of mindfulness and some of the positive benefits to self-esteem and behaviour result.

Ultimately, by being enabled to take responsibility for her own actions and choices of how she responded to Duchess's behaviour, Cinderella may have been experiencing aspects of the experiential freedom and organismic trusting proposed by Shaver *et al.* (2007). By being able to reach a mindfully embodied state she was able to act in an authentic manner. When Cinderella initially approached Duchess her body language reflected her inner emotional state of mind – of being angry at perceived criticism – so the mare reacted accordingly by rejecting her. Cinderella's normal pattern of behaviour in response to rejection was to become increasingly angry, but in this case she was in an environment where she was able to practice a different approach. Being encouraged to bring a sense of awareness to her actions without negative

168 *Equine-Assisted Therapy and Learning with At-Risk Young People*

judgement and within a peaceful environment may have enabled her to relax, which would appear to be a crucial part of all mindfulness practices, Coleman writing that 'being outdoors provides mental space and clarity, allowing our body to relax'(2006: xv).

Some further observations on horses and mindfulness

It may not have been possible for many of the young people who attended The Yard to consciously process some of the suggestions or practices put forward in the mindfulness literature, since the level of self-awareness at adolescence is different from that in adulthood, and their own particular difficulties surrounding their traumatic pasts were often too raw at the time they participated in TH. However, many of the elements involved in working with horses would appear to embody a number of the mindfulness techniques described by Segal, Williams and Teasdale (2002) and Zylowska *et al.* (2008), amongst others. Consequently, young people participating in TH may be enabled to achieve some of the same benefits that mindfulness practices claim, such as reaching a state of calmness and heightened awareness. This is linked to heightened self-knowledge and hence self-regulation according to Brown and Ryan (2003) and Chambers, Gullone and Allen (2009). Mindful experiences with horses may be more accessible for some at-risk young people with behavioural, emotional or attention challenges than traditional MBSR or MBCT or other therapeutic techniques. This was demonstrated through the examples with Cinderella, Wayne and Freya, who all appeared to gain some benefit from participating in exercises and techniques with the horses which had some similarities to those employed in some mindfulness practices. Perhaps a more important and useful application of TH, however, is that the benefits suggested to be gained from mindfulness practices, in particular MBSR, are achieved naturally by working with horses, as these same principles of relaxed awareness are a requirement of a successful partnership with the horse; and the young people, almost without exception, were motivated in wanting to achieve this relationship.

Psychosocial and psychospiritual dimensions of horses and nature

The benefit of being in the natural environment was a thread running through several of the other themes that arose through the research study, and it has various layers of meaning. Therefore it has been difficult to separate the effect of being in nature from some of the other themes. Being in nature was a backdrop to the TH sessions, and is referred to as a 'background to activities' by Wals (1994). At the same time, the natural environment had its own characteristics which were considered to be fundamental to the whole experience provided at The Yard. Some of the themes that emerged were articulated by both the young people and the adults, others just by a number of the adults who went into some depth with theoretical and philosophical aspects related to being in the natural environment. As was the case with many of the other themes, the young people would often speak about their experiences at The Yard through their perceptions of how the horses felt about certain things. This was the case with the theme of 'being calm', which was generally attributed to factors associated with the horses for the young people but was more strongly identified with the natural environment for the adults. Whilst 'feeling calm' has also been explored previously in relation to mindfulness and psychotherapeutic insights, this section looks at how 'being calm' also appeared to be linked to being in the natural environment. Some of the nature and ecotherapy literature talks of the spiritual dimensions of being in nature, and of how it can be 'restorative' in its own right (Kaplan, 1995). Although the data are more limited in this area due to methodological difficulties involved in gathering aspects of the spiritual dimensions of being in the natural environment from the young people, Cinderella did, nevertheless, write that 'being in nature de-stresses you' in a questionnaire about her experiences of being at The Yard. Mostly, however, it was not through words but through their body language, attitude and behaviour that the participants would appear to relax and engage once we got out into the fields and open spaces. Another area that both the adults and some of the young people talked about was of how the environment and being with horses was linked to 'feeling free' and 'escape' in some way. These concepts draw on theory from both the nature and the ecotherapy literature, and some of the Jungian theories around symbolism and archetypes.

Lastly there are the more obvious benefits of physical activity involved in working with and caring for horses, and how this was possibly linked to the young people being more willing to engage in educational and learning activities. Again, as these are not things the participants spoke about directly, data from fieldnotes and from interviews with the adults are drawn on in more depth to describe these elements of TH that emerged from the research.

Nature, horses and 'being calm': The natural 'restorative' environment

The view of the importance of the environment in facilitating the young people to relax, and be more open and motivated to learning, was voiced by many of the adults involved in the young people's lives. There is a large body of literature claiming that the natural environment is fundamental to health, and that our distancing and dominance over nature is largely responsible for many elements of mental, physical and spiritual ill health (Kahn & Kellert, 2002; Louv, 2008; Nebbe, 2000; Roszak, Gomes & Kanner, 1995). The deep ecology movement goes further, suggesting that we have a 'natural bond' with the biosphere as we are all interconnected (Lovelock, 1979). Other literature makes links to the particular psychological benefits of the natural environment, with Kaplan (1995) talking of the 'restorative environment'. Berger and Mcleod state that 'the concept of conducting transformative and healing work in nature is not new' (2006: 80). They and other authors claim that the healing power of nature can be traced back historically to when humans lived alongside nature and animals, and they understood, in a lived, embodied way, the connection of all things (Kahn & Kellert, 2002; Roszak, Gomes & Kanner, 1995).

This understanding of the natural environment as being an important element of TH for the participants was expressed by the adults in different ways. As previously shown, many of the adults involved in the young people's care spoke of how they would be calmer after attending TH sessions. This was repeated by the foster carer, Laura, writing about another participant in a feedback form, '[He] is always calm and relaxed after a session.' Whilst most of the data concerning possible links between nature, being calm and a therapeutic environment were observational in terms of the young people, many of the participants did comment on how the horses helped them to feel calmer. In a questionnaire that they completed for the research, Cinderella and Minimax both wrote that horses made them feel 'calm'. When asked what it was that she especially liked about Ruby – as she often chose to work with her on TH sessions – the participant Freya replied that it was because 'She looks sort of kind...and calm.' And during a conversation with Cinderella in one of the sessions as we chatted about the horses, she stated that, in addition to the horses, being in nature was important to her in terms of destressing: 'It's like, if you're stressed, being in nature de-stresses you.'

Horses, Mindfulness and the Natural Environment   171

Both the counsellor Sally and the therapist Deana talked about the importance of being in the natural environment and of how it contributed to the therapeutic process. Sally explained that she believed that there were a number of reasons why being in nature was an important element of EAT/L, including how the horses can be instrumental in helping everything to 'slow down'.

> And that stillness, I mean, if we weren't talking now we'd just be sitting here quite quietly wouldn't we? The horses have just been pottering around quite quietly, and the whole thing just slows down.
>
> (Sally, counsellor)

She went on:

> And to be in an environment like this with beautiful views, it's very still, it's got bird song, and you know, there is something very elemental about being outdoors because you've got the air, the rain, the earth, you know, everything around you. And the building removes you from that, the kind of elemental nature of being outside.
>
> (Sally, counsellor)

Deana reiterated similar claims, also mentioning bird song and views in her understanding of the importance of the natural environment in the therapeutic process, stating:

> I think this environment is very special and you can hear the bird's songs, the spaciousness that you can have here and just being able to look out across nature. It feels really holding as well. Yeah, so I think the environment has a lot to do with it.
>
> (Deana, therapist)

As mentioned previously, it is claimed that the very nature of therapy can create anxiety for some young people (Biegel *et al.*, 2009) and that introducing contact with animals within the therapeutic environment can help to provide a calming effect, especially during initial sessions (Levinson, 1969; Walsh, 2009). One reason why horses may be effective in initiating a calm atmosphere is that they inherently want a calm and peaceful existence as they need to conserve energy for when it may be required to escape potential predators. In addition the horse will always look for a calm and intelligent leader that it can trust. Therefore, in

172   *Equine-Assisted Therapy and Learning with At-Risk Young People*

order to have a successful relationship with a horse, it is necessary to behave in a calm manner. Some young people may forge relationships with other animals that also require a similar, calm approach. Talking about her work with a 15-year-old boy and his beginning to establish a relationship with a llama, the psychotherapist Brooks says how llamas, like horses, are curious and will want to approach a person who remains calm and unaggressive (Brooks, 2006). This is the case with many of the animals found in the outdoor, natural environment as opposed to domesticated pets found in the home, so these animals demand an additional awareness of body language and emotion regulation. At The Yard it was often found that the young people would become transfixed by the rare sight of a wild fox or deer in the surrounding fields and woods surrounding and be able to monitor their body language accordingly, becoming very quiet and calm.

'Horses for courses': Horse management and a calm, relaxing environment

An important factor inherent in the philosophy and practice of The Yard was of creating an informal and relaxed atmosphere for both the horses and the young people. The natural environment, horse management and person-centred orientation of The Yard were all interwoven and equally important in helping to create a 'therapeutic environment' and 'safe space'.

Talking about both the environment and how the horses were managed within a natural horsemanship system and philosophy, Deana explained:

> Well, it might be that they're closer to their natural state. That they live with the herd, and that they have much more freedom... they are... um they live outside... not shut in stables. Yes of course they're domesticated, but for me they are... there's a respect for their... um the horses' spirits. [They're] respected for who they are.
>
> (Deana, therapist)

The way in which the horses were kept was something that seemed important for the participant Emma, who explained how the pony Timmy was much calmer at The Yard than he had been at his previous stables, which adopted a more conventional approach where the horses were kept in individual stables and separate paddocks. Emma

Horses, Mindfulness and the Natural Environment   173

went on to say that he had been permanently anxious and unsettled at this stable yard and how she thought all of the horses were calmer and more 'settled' at The Yard than at other stables she had been to. When asked why she thought this was, she said:

> Well, here they're not forced to go in the stables, are they? They can come and go as they please...And then they've got really big fields and they're not separated.
>
> (Emma, participant)

Linda, Emma's adoptive mother, reiterated how she felt that the general environment of The Yard was helpful to Emma in terms of suiting her particular needs. She explained:

> I think it's a sort of more comfortable environment, so she feels more relaxed...She didn't like the pressure of routine at [her last stables]. You know, when you have to get a horse brought in and out at a particular time, feed them at this time, you've got to do this, you've got to have it [horse] turned out with that one and you've got to have it there, and you know, it was very regimented at the last place. And I think that was a difficult thing for her. She really does not cope with, you know, strict regime, you know. She struggles at school in that respect.
>
> (Linda, Emma's adoptive mother)

By creating a relaxed environment for the horses it was hoped that this would assist in creating a relaxing environment for the participants. This is something that the EAT/L practitioner and counsellor Sally reiterated. She explained that, for her, the impact of the environment on the horses' well-being is fundamental in creating a calm atmosphere, which she believes is vital in terms of creating an effective therapeutic environment, adding: 'the horses are relaxed too in this environment, and that's crucial'.

Just being outside and relaxing with the horses was something that a number of the young people seemed to want to do and benefit from. Accordingly, The Yard tried to facilitate and encourage this whenever possible, a non-directive approach being able to accommodate a flexible approach to TH sessions. On one occasion in a session with Minimax, he chose to spend time just lying down in the field with the mare Ruby

174 *Equine-Assisted Therapy and Learning with At-Risk Young People*

and he was left to relax with her whilst we busied ourselves on some field-management tasks nearby.

Minimax and Ruby relaxing together

It was a lovely warm sunny summer's day today with a bright blue sky and a warm breeze on which the birds were drifting across the valley. When we went out into the field to bring in Ruby and Jason, Ruby was lying down, quietly dozing in the sun. Minimax went straight over and sat down on the grass next to her; he seemed very relaxed with her, as did Ruby with Minimax. She was quite happy to remain lying down with him – which is quite unusual for horses as they are obviously extremely vulnerable lying down and will only remain so with someone they completely trust and feel secure with. We pointed this out to Minimax, who shrugged – but with a grin on his face as though he knew the two of them had a special relationship. As we carried on with pulling up ragwort (poisonous plant to horses), Minimax and Ruby remained lying down together, the mare with her eyes half closed, Minimax quietly relaxing next to her, seeming to be lost in thought.

(fieldnotes)

In a study with young adolescents from urban Detroit, Wals found that several of the participants spoke of how nature was a place 'where you can be alone and think by yourself' (1994: 19). The author states that

away from their everyday distractions, these students can think freely about their emotions. The peacefulness, the slow pace, the sounds of the birds, the sun shining through the leaves, are some of the elements of nature that, to some of the students, are soothing and inspire reflective thought.

(Wals, 1994: 19)

Often Wayne would also seem to just want to be on his own with the horses and would spend time quietly stroking and hugging Ruby. One day we were all sitting at the bench in the yard and I was trying to ask Wayne questions about the horses for the research, not very successfully. He seemed to find this distracting and asked if he could go and eat his lunch with the horses instead. He then went and put his chair next to the fence amongst the trees so that he could watch the horses whilst he ate his sandwiches. It was a very peaceful, warm spring day, the birds were singing, flowers were out and the only sound was the occasional snort of a horse or a sheep bleating in the distance. Altogether it was very relaxing. After he had finished his lunch, Wayne came back and said: 'I could just sit here and watch the horses all day.' Although he had not commented specifically on the environment, it certainly appeared that it was a combination of the factors of the horses and the natural environment which had seemed to attract him to wanting to spend time just sitting quietly contemplating them. A more conventional yard of metal and concrete where the horses were shut in stables, with the associated stress-related behavioural problems of kicking doors and biting, would possibly not have created the peaceful environment which seemed to engage Wayne on this occasion. Louv (2008) and Taylor, Kuo and Sullivan (2001) make the claim that being in nature can alleviate the symptoms of ADHD in young people due to its 'attention restoration' ability. It would appear that Wayne was demonstrating how being in nature with animals can create a relaxing, yet motivating and 'attention restoring', environment.

'Special places', learning opportunities and the natural environment

One sunny summer's day after he had been attending TH for a number of months, Minimax came bounding into the yard, exclaiming 'I love being here!' This was quite an expressive statement for him, who normally kept his conversations directed towards talking about the horses, apart from the occasional outburst about his brother and his thoughts on social workers (not very positive). When quizzed a little bit more about what he meant by this, Minimax characteristically clammed up again, just muttering, 'I don't know, just everything', and busied himself with changing his boots. As he appeared not to want to expand on his comment, I didn't question him any further. This is an example of one of numerous times I was faced with one of the many ethical dilemmas inherent in being a practitioner-researcher – of wanting to persist in questioning in the quest for more information but, at the same time, being aware of the frustrations and limitations in this endeavour. Ultimately, as a practitioner first and foremost, the young person's well-being took priority, so there was a clear ethical distinction for me. However, I was aware that there were times when further rich data could perhaps have been generated by further questions.

Some authors have written about the importance of favourite or special places to both adults and children, and especially natural places, believing that these have an importance for a number of reasons ranging from cognitive and developmental (Heerwagen & Orians, 2002; Kellert, 2002), emotion regulation and restoration (Korpela *et al.*, 2001), attention restoration (Kaplan, 1995), attachment to place (Jack, 2010) and as a therapeutic 'sacred' space (Berger & McLeod, 2006). Whilst the young people who attended The Yard hadn't chosen the space themselves because they had been referred for TH, nevertheless they may have benefited from some of the aspects that these authors refer to, as demonstrated by Minimax exclaiming that it was 'everything' about The Yard. Wals (1994) suggests that the natural environment is important for young people to escape some of the problems in their home environment, and in how it 'provides a friendly atmosphere away from the worries many of the students have at home or at school' (1994: 16). He also speaks of how nature is intriguing, how 'nature piques the curiosity of these students and in doing so inspires learning' (Wals, 1994: 18). This was something that appeared to be the case with two of the young people, Lucy and Freya, who both showed an especial interest in nature and the natural environment. Lucy had a good knowledge of wild plants, which she had learnt from her school, and whilst we were out in the fields or out hacking in the lanes she would often point out some that she knew were edible. She also knew the names of lots of wild flowers and animatedly told me about the importance of flowers and bees to ecology through pollination. During one TH session with Freya we were walking around the fields conducting a routine field-maintenance check after which the young people then sometimes completed a field-maintenance quiz. On this occasion the wild garlic had grown up profusely in the small woodland at the bottom of one of the fields. We showed Freya how these could be picked and eaten, demonstrating this by eating some ourselves. At first Freya feigned disgust, saying: 'Uuugh, I'm not eating that.' However, she was interested in participating in identifying the correct leaves and picking some of the younger tender ones. Afterwards we took some back to the yard and Freya took a few leaves with her when the staff member from her residential home came to collect her. We were then surprised and pleased when the following week she asked if we could go and pick some more of the wild garlic to take back to put in a salad for dinner that evening. The staff member commented that normally Freya wouldn't eat greens in any form so it seemed that somehow having a connection to the plant and process by picking it herself had facilitated a positive change in her eating behaviour.

This interest in plants and the natural environment accords with biophilia theory, which claims that we have a natural fascination in and affinity with our natural surroundings due to our long history as hunter-gatherers (Kellert & Wilson, 1993; Wilson, 1984). Whatever the reasons behind Freya's interest in nature, the experience of identifying and picking the wild garlic provided an opportunity for her to participate in a valuable learning experience as suggested by Wals (1994), when he talks about how nature inspires learning. In her study of the effects of 'greenness' on children in urban environments, Wells goes further to suggest that engaging in the natural environment has been shown to 'have a profound effect on children's cognitive functioning' (2000: 790). Katcher and Wilkins (2000) suggest that through being motivated and inspired to learn about nature, this then transfers to enabling communication skills and for dialogue to be opened. They write: 'learning came from nature, but it came into a dialogue between people. Learning was a social skill and was designed to equip a child to enter into conversation with others' (Katcher & Wilkins, 2000: 165). For some of the young people, such as Freya, who because of their negative experiences could find it difficult to engage with adults who they had learnt not to trust – often with subsequent knock-on effects on their education – finding an environment in which they could engage with adults and find opportunities to learn and communicate could be the beginning of building more positive opportunities and futures.

Horses, nature and relationships

Another idea which has resonance with EAT/L is that each part of nature – the landscape, the elements, the weather, animals – can 'trigger parallel psychological and spiritual quests that can open a channel for mind-body work' and that nature can act as 'a bridge between people' (Berger & McLeod, 2006: 87). Talking about a difficult time in her childhood, Lucy explained how the pony Minnie had brought elements of this to her life, helping her to build relationships with others:

> Lucy: Because I already had a good repertoire [sic] with animals, what I needed was a repertoire [sic] with humans as well. She was...she was the table.
> HB: Like a bridge?
> Lucy: Yeah. Yeah.

(I understood Lucy to mean 'rapport' when she talked about 'repertoire'. She expanded in later discussion that she was trying to get across that what she needed was to learn a way of communicating with humans as well as animals).

Emma's adoptive mother, Linda, talked about how she felt that the environment, in terms of being outdoors, the orientation and practice of The Yard, and with the horses as the motivating factor, had helped her and Emma to build on their relationship, which she explained had sometimes been difficult. Linda said that Emma struggled in the school environment because she found the structure and social side of it really difficult, explaining: 'It's something she just lacks completely.' She went on to say how difficult it was to motivate Emma to do anything, both in school and at home in the holidays, but, because she 'likes being outdoors' and due to her interest in horses, she found that this was something that she could encourage. This was in the hope that it would benefit Emma, in terms of both physical health and social skills, and at the same time help with their relationship, explaining:

> It's been a good experience for both of us you know. And it's really been a positive thing for us building a relationship. Not always having to battle. So the experience I feel now, is hugely better and more balanced than before. And I mean, on the way there and on the way back home in the car we're talking, so even that part of the, um, you know, going to see the horse is positive because we'll have a chat on the way there and a chat on the way back which is positive, instead of her just sitting there in silence or getting out of the car and walking off.
>
> (Linda, Emma's adoptive mother)

This appeared to correspond with what Emma said when asked about the differences she felt between being at The Yard and being in other :

> Yeah, so like after I've done everything and I know I'm going to be going home soon I just suddenly get really bored and I don't know why. But I mean like I get really excited on the way there, but on the way back I'm like: 'OK, I'm bored and I don't have a *clue* what I'm going to do when I get home.'
>
> (Emma, participant)

The counsellor Sally talked about how she had a mixture of animals at her EAT/L practice, ranging from chickens to rabbits, guinea pigs and dogs in addition to a horse, as she believed that this was useful in facilitating interest and conversation through the interactions between the species. She said that this 'provides lots of opportunities to talk about things' and that the horses and other animals

> in a sense help you get back in contact with something... it's not that you don't know it, it's just somehow... it got buried.
>
> (Sally, counsellor)

On the larger farm on which The Yard was located there were sheep and lambs, chickens, a rather mischievous goat and a rich abundance of wildlife, such as many varieties of birds, together with rabbits, foxes and the occasional deer. This provided useful opportunities for discussion and quiet contemplation. An example of this was when during a TH session with Kelly a fox padded silently across one of the fields with a rabbit in its mouth, causing Kelly to sit down on the grass with us and watch with fascination as it crossed our path, seemingly unconcerned about our presence as long as we remained completely still. At times like these, it felt as though we were all connected in some way to the moment that was unfolding and to the other animals and larger environment in a way that is difficult to express in words.

Connection and 'listening to your instincts'

The idea of being in contact with something else that Sally mentions is perhaps linked to the idea of being connected to a larger entity in the way that the Gaia hypothesis and ecologists speak of everything being connected (Lovelock, 1979). Nebbe states that 'humans evolved with the earth, with and within the natural environment' and that we therefore have a basic need to connect with the natural environment (2000: 391). In their study of therapeutic horticulture, Sempkit, Aldridge and Becker (2005) found that participants spoke of feeling free, connection to the natural environment and how they found distractions from their problems, all themes which arose in the current study and are perhaps related to being in the natural environment alongside the interactions with the horses. The gestalt ecotherapist Cahalan (1995) writes about 'ecological groundedness', which he suggests is a state of healthy functioning that can be gained by connecting with nature. Perhaps we experienced elements of this in the TH session with Kelly whilst we watched the fox.

180 *Equine-Assisted Therapy and Learning with At-Risk Young People*

Deana spoke of how she believed that the horses were part of the whole environment and therapeutic process when she said:

> The recognition that we are connected to each other as the horses are connected to each other as well, um. And I think that has a profound effect on young people.

(Deana, therapist)

The participant Lucy spoke of how she felt that it was important to listen to your instincts when around horses, and of how city life could repress this:

> Lucy: ... and you're not sensitive to horses' or animals' needs because you're too wrapped up in, you know, city life and your proper senses haven't built or whatever, and you're just going to make the situation worse. So if you listened in the first place, listen to your heart rather than your head.
>
> HB: Right. Right. So when you're around horses, it's more important to listen to your, your ...
>
> Lucy: Your instinct

Freedom, escape and getting away from problems

This idea that being in nature, in the elements and with horses and other animals can reconnect us to something is also perhaps linked to some of the things that both the young people and the adults talked about, of how being with the horses in the natural environment took the young people away from their problems and gave them escape from the stresses of their other lives in some way. Lucy explained that for her, being with horses and animals in nature helped her 'to escape and keep busy. Escapism.' She went on to talk about how coming up to be with the horses 'it kind of, it kind of helped me focus on something else rather than me'. When she was asked if she could expand on what she felt was more important for her, the horses or the environment, she answered:

> Well, I'm not sure. I think it's a bit of both with me personally because the thought of horses ... animals, makes me calmer anyway ... Um, then, just kind of the thought of going to see the horses makes me kind of push out any irritations.

(Lucy, participant)

Horses, Mindfulness and the Natural Environment 181

Korpela *et al.* (2001) in their study with adolescents found that they spoke about their favourite places in nature as places they went to calm down and escape from their problems. Another view proposed is that being in nature is restorative, Kaplan (1995) calling this 'attention restoration theory'. This may be relevant to what Lucy refers to, and to what Freya experienced in her interest in the natural environment. Some of the adults also brought up what they perceived as the importance of being with the horses and the environment as helping the young people to get away from their problems. Angie stated that after TH sessions at The Yard, Minimax would be

> calmer, it seems to take their minds somewhere else, gives them a different view on things, it takes them away from the problems...yes he would talk about the horses instead of that.
>
> (Angie, foster carer)

Deana too, spoke of how different she would feel when she was at The Yard, saying:

> You know, how different I feel when I'm up here, in myself. I can breathe! Like a weight has been lifted off my shoulders. Yeah. And I stop thinking about all the things that I worry about that wake me up at four o'clock in the morning. You're just with the horse, smelling, feeling, riding.
>
> (Deana, therapist)

Talking about one of her clients with Asperger syndrome who had limited verbal skills and would often appear anxious and frustrated at the residential facility where he lived, Sally described how his behaviour and general demeanour would change once he was at her EAT/L smallholding:

> So the minute he has space around him and fresh air, he doesn't feel cooped up. He feels he can move freely and that has a big impact on his behaviour and on how he's feeling, his well-being. And that is about this isn't it? Being outside, in the elements, and just kind of feeling more free to be.
>
> (Sally, counsellor)

For Emma, being at The Yard seemed to signify freedom on different levels. She explained that she liked being at The Yard more than school because [school] is like really boring and crowded and stuff, this place feels like it's got more freedom and stuff.

(Emma, participant)

Emma went on to say that she envied horses and that she was 'jealous of them' as they could do what they wanted and didn't have to worry about behaving in certain, socially acceptable, ways. This was because

In the wild they go where they want, when they want, they go splashing in puddles and not worrying about getting wet, roll in mud.

(Emma, participant)

Jung writes about horses in dream interpretation as representing instinctive drives and as a symbol of the individual wanting to break loose from restricting conscious control, seeing them as a positive message in this instance (1978). In addition to Emma talking about how horses represented freedom in some respects, Lucy may also have been referring to aspects of this when she talked of how horses could help you to connect to 'your instinct'.

Horses, myth and magic

These themes of how horses represent freedom and wildness on some level are perhaps linked to cultural aspects; of how the horse has been incorporated and represented in culture throughout history as often symbolising different parts of our psyche related to themes such as power, domination and our instinctive drives, together with their role in magic, myth and dream symbolism (Atwood, 1984; Howey, 2002; Jackson, 2006; Jung, 1978). An example of how the horse came into myth and legend in the daily life of The Yard is given by the mare Duchess and Cinderella. One day during a TH session with Cinderella, Duchess came in from the field with a strange purple mark across her wither (top of shoulder) which we thought was probably bird droppings from a bird that had been eating purple berries. However, before we had the opportunity to voice this deduction, Cinderella cried: 'Look, she has been touched by unicorn dust!' Many of the young people would call Duchess 'the fairy horse' due to her pure white coat and her wise and gentle nature, and would often initially be drawn to her before gaining the confidence to work with the other younger, or more challenging, horses. This example serves to illustrate how children's imaginations and stories are interwoven in their everyday lived experiences, and how the horses could precipitate this side of their imaginations. In Jungian psychology the unconscious and imagination are important aspects in the quest for self-understanding, and, although Jung wrote about these in terms of adults, the role of fairy tales in child development is suggested as providing a culturally socialising function (Jones, 2008). White horses, such as unicorns and Pegasus, have long been associated with being healers, wise or good helpers to humans in some way; they are archetypal in that respect (Held, 2006; Jackson, 2006). This theory also perhaps has connections to the 'collective unconscious' that Jung referred to (1983, 1990) and how white horses have become embedded into a collective understanding of their significance. An interpretation of Jung by Hannah, Frantz and Wintrode (1992) goes on to suggest that horses can show us the way when we are lost and cannot help ourselves, due to their clairvoyant and divinatory power.

Another area where horses appeared to enable some of the participants to connect with other aspects of themselves was in the way that they empowered the young people to feel more powerful and confident, such as when Lucy exclaimed: 'I feel like the queen of the world' when she was riding. There seemed to be connections between being in the natural environment with the horses, which led to wanting to try new things, to be open to learning and to taking some risks. Nebbe (2000) describes how more positive ways of being can be gained by being in nature. She suggests that the very nature of being in the natural environment with exposure to the elements brings us closer to our ancestral roots of having to adapt and learn certain skills and coping behaviours in order to survive, giving examples of these coping behaviours as 'self-sufficiency, risk-taking, initiative and co-operation' (Nebbe, 2000: 394). Some of the young people spoke of how they especially liked the horse Jason because he was a 'challenge', and would want to try new and more challenging activities on the horses. An example of this was when Kelly stated wistfully one day when she was riding Ruby along the lane next to a large, sloping, inviting field during a TH session: 'I wish I could just go galloping off across that field'.

Physical activity and learning in the natural environment

Linked to the aspect of risk-taking inherent in being involved with horses that Atwood (1984) and Scanlan (2001) suggest as being a strong motivating factor in their attraction is the physical activity involved in looking after them. It is well documented that exercising is linked to

184 *Equine-Assisted Therapy and Learning with At-Risk Young People*

feeling good through the release of endorphins (Halliwell, 2005). Mucking out stables, maintaining the fields, throwing up the muck heap, changing and carrying water buckets, filling and hanging haynets – these are things that need doing daily in addition to grooming and exercising the horses. Many of the young people who attended The Yard participated in little physical activity, and their carers would voice despair regarding how difficult it was to persuade them to engage in outdoor pursuits, such as joining them for walks, or taking up a sport or other healthy outdoor activity. A number of the participants were overweight and unfit, and some had food issues, although others, such as Minimax, were sometimes hyperactive. Whilst, on occasion, some of the participants would seek to avoid certain tasks that they particularly disliked – for some this was filling haynets, for others it was mucking out – generally all of them would engage in these tasks once they understood that they were necessary for the horses' well-being and health. Alongside the inherent physical aspect of looking after the horses, educational opportunities were created wherever possible within the context of learning about horses. This could be learning the 'points of the horse', which involved reading from a diagram and matching it against the real horse, to measuring and weighing their feed. TH sessions could involve more, or less, formal educational opportunities depending on the requirements of the participant and referrer, and included projects on the role of the horse through history to learning about the horse's food production and links to the natural environment. The following extract is from fieldnotes about a meeting with a staff member from Wayne's PRU.

Learning in context

Today a staff member from the PRU which Wayne attended came out to The Yard for a meeting. This was to discuss some of his learning goals and possible curriculum work we could incorporate into sessions with the horses. She immediately commented on what a peaceful, relaxing environment she found The Yard and how she believed it was more beneficial for Wayne than the school environment in which he struggled. We went on to discuss Wayne and his academic level and what sort of things we could introduce into TH sessions of an educational element. She explained Wayne's educational ability as being at low primary level, which surprised me because he had managed reasonably well with the points of the horse sheet we had given him, although he had obviously struggled with unfamiliar words such as 'wither', 'stifle' and 'fetlock'. When she sees this sheet she says she is surprised that Wayne had even attempted the worksheet due to the size and amount of writing on it and level of difficulty of words. She claims that Wayne would normally have played up and refused to attempt something of this level: 'I'm surprised he had a go at that. Normally at the centre he would have started playing up if we had given him something that difficult... or walked out or something.' She goes on to suggest that perhaps this is due to the combination of being in the outdoor, natural, environment, together with the educational tasks that have meaning because they are related to learning about something he is motivated about and so have context.

(fieldnotes)

Emma's adoptive mother talked about how she believed that Emma's motivation to be with horses and her preference for being outdoors, as opposed to being inside school, had helped her with physical aspects in addition to learning and social skills. She described how she had huge issues around food, and Emma herself stated: 'I will just chuck food around and stuff'. By being motivated to improve her riding and be able to progress onto bigger and fitter horses, Emma was able to understand that she too needed to have a balanced diet and exercise:

The thing with horse-riding that is eventually going to make her get fitter and eat better, you know, is that she's got to eat properly because she's got to be fit, she's got to have exercise if she's going to be a good rider. You know, it's almost sort of lurking in the back of her mind. It's motivation.

(Linda, Emma's adoptive mother)

An aim for Cinderella on her referral form was for her to engage in more physical exercise. In addition her respite carers explained that it was difficult to persuade her to eat a healthy diet. Food was an issue that often came up during TH sessions with Cinderella, who would want to give the pony Leo extra portions in his feed. In a questionnaire that she completed for the research she wrote about Leo: 'I gave [him]) a little bit extra today 'coz he's special!' However, unfortunately Leo suffered a painful foot condition (laminitis) resulting from eating too much rich grass and getting overweight in his previous home, and therefore needed to be on a restricted diet together with managed exercise. During the following TH sessions, Cinderella learnt about the importance of a

186 *Equine-Assisted Therapy and Learning with At-Risk Young People*

balanced diet, of the importance of vitamins and minerals, and of the need for exercise in order for Leo to remain healthy, and it was hoped that this knowledge would be transferable on some level to her own health and well-being. Certainly she showed that she cared about Leo's well-being by being willing and happy to walk him out in hand for his exercise regime, which, in turn, helped with her own exercise.

Space to let off steam

For some of the other participants, being outside with space around them could be an opportunity to let off steam. One day Minimax was accompanied by another boy on a temporary foster placement with Angie. This boy, like Minimax, was quite exuberant, which, together with Minimax's difficulty with his peer group, ensured that it quite quickly became clear that it was unsafe for them to be together in the yard with the horses loose. This was because the horses were picking up on their agitated behaviour, which was causing them to become anxious and agitated themselves.

Working outside with the horses

Deciding it would be better to get the boys outside into the fields and leave the horses to eat their hay in peace we suggested a field maintenance session, telling the two boys that we needed to check all the fences and also do a general safety check of the fields. We explained that we needed to look for poisonous plants and any broken branches and rubbish that may have blown in after the storm of the night before. We left them to work out how they wanted to do this to themselves but gave them the suggestion that they could work together or separately depending on what they decided. They decided to complete the task by running up and down the steep hills, looking at the fences and calling suggestions to each other whilst successfully wearing themselves out – to our relief! Once they had calmed down it was possible to gain the boys attention and show them how to push the electric fencing posts back into the ground with their boots. We talked about the reasons why it was important to divide the field up so that one part was rested over the winter – in order that the grass could regrow in the spring and the parasitic horse worms die out. Once we had completed this task we continued around the field doing a field safety check and encouraging the two boys to identify different plants which could be both good and bad for horses to eat. We also looked for examples of safe and unsafe fencing and areas the horses may find useful for shade or shelter, such as under trees or hedges, in order to protect themselves from the weather. Afterwards, sitting at the bench in the yard with a cold drink, they both completed the field maintenance quiz, although choosing to do this individually in quiet concentration. Following this session, Minimax's carer, Angie, told me how Minimax had been engaging in particularly risk-taking behaviour recently, climbing up very high walls and trees, and putting himself at risk. We wondered if by exploring safety, within the context of keeping the horses safe, this might have transferred to aspects of his own safety on some level.

(fieldnotes)

This extract serves to illustrate how the environment could be useful for both aspects of fitness and exercise; in the case of Minimax and the other boy to let off steam, and, for other participants, such as Cinderella, to encourage more physical exercise. A study by Taylor, Kuo and Sullivan (2001) shows how young people diagnosed with attention deficit disorder (ADD) were better able to concentrate and engage in both school work and other activities when in 'green settings'. They suggest that 'before engaging in attentionally demanding tasks, such as schoolwork and homework, ADD children might maximize their attentional capacity by spending time in green settings' (Taylor, Kuo & Sullivan, 2001: 73).

It is increasingly being recognised that childhood obesity is increasing and is linked to health problems (British Medical Association, 2005; Ebberling, Pawlak & Ludwig, 2002; McCurdy *et al.*, 2010). Munoz writes that 'it has been suggested that those with access to natural outdoor areas, that they can use easily and feel comfortable in, have higher levels of physical activity' (2009: 6). Whilst it was not necessarily particularly easy for the young people to access TH as this required a referral, an objective of The Yard was to create a relaxed and 'comfortable' environment so this may have helped facilitate physical activity in the way that Munoz suggests. Other authors have written about how young people from deprived backgrounds are particularly excluded from the natural environment, and are at greater risk of obesity and related health problems (Ward Thompson, Tavlou & Roe, 2006; Wells, 2000). As the participants who attended The Yard were often from these categories, it could be argued to be especially important to find motivating engaging outdoor activities, such as those provided by EAT/L and TH, for these groups of young people.

188 *Equine-Assisted Therapy and Learning with At-Risk Young People*

Conclusion

This chapter looks at how being with the horses and the associated exercises and approach of The Yard have similarities to aspects described in the mindfulness literature. A major theme running through the research was related to calmness, both of the young people's own accounts of feeling calmer when with the horses, and the adults reporting their behaviour as being calmer after attending TH sessions. Often the young people talked of how it was important to be calm around the horses in order to be a calming influence on them and to help them to feel safe. In order to achieve relaxation and more mindful behaviour, some mindfulness practices, such as MBSR, which employ exercises such as the body scan and of being in the 'present moment', have many similarities to aspects of being with horses, and some of the emphasis and exercises employed at The Yard. In addition the chapter explores how relating to the horses and learning how their behaviour could influence them could bring a growth in self-awareness, and perhaps has links to the 'self-regulation' and 'authentic functioning' that the mindfulness literature refers to.

Another key area was the role of the natural environment within the context of TH at The Yard. A central tenet of the philosophy of The Yard was that being in nature offers therapeutic benefits due to a number of practical and theoretical reasons. It is also related to the style of 'natural horsemanship' practised, which strives to manage horses within an environment that is as 'natural' as possible and as near as possible to their wild state. Practically, being in the natural rural environment offers 'space to let off steam'. It provided an opportunity to run around and be noisy for some young people, many of whom found the formal classroom, or therapy, environment difficult. Once they had expended some energy it seemed to enable some participants to be more motivated to learn and engage with the practitioners and horses. The physical activity involved in working with horses was also a suggested 'spin-off' benefit from TH sessions, with an understanding of the relationship between physical and mental health being noted. On a philosophical level, theories drawn from some of the environmental literature seemed to be applicable. Themes of the motivating aspects of the natural environment and of being drawn towards being in nature from biophilia theory may be relevant, as illustrated by Freya being interested in foraging for naturally growing garlic. Other relevant theoretical literature speaks of the 'restorative' benefits of nature and of it being a place where young people can escape from their problems and be calm. Conversely, the natural environment can be a wild place and some of the young people found how the weather and changing seasons can affect the horses, as demonstrated by the TH session with Wayne and Ruby on a windy day. This was connected to suggestions of how the natural environment and horses are culturally linked to themes of freedom and wildness. Some young people appeared to find the environment of The Yard somewhere where they could explore risk, and overcome and set themselves challenges, such as when Kelly wished she could go 'galloping across the fields', and Lucy talked of how she liked a 'forward going' more challenging horse. It would seem that the natural environment perhaps offers what Wals (1994) terms a 'background to activities' in which all of the other aspects fitted, and that these were all ultimately interlinked so that one could not occur without the other, in a similar way to the Gaya hypothesis (Lovelock, 1979).

7
Conclusion

Introduction

This book arose out of my experience of being a practitioner at The Yard, a TH programme that I established for a foster care company in the UK. As a qualified social worker with many years' experience of working with and training horses, together with additional training in EAT/L, I saw an opportunity to provide an alternative learning and therapeutic experience for at-risk young people through the medium of horses. However, despite much anecdotal support for the benefits of EAT/L, there remains limited research in the field due, in part, to its being a relatively recent addition to the therapeutic domain, and also because of the financial, political and practical limitations inherent in research into alternative therapeutic approaches (Bergin & Garfield, 1994; Daniel-McKeigue, 2006; Selby & Smith-Osborne, 2013). Due to the additional ethical and gatekeeper problems involved in undertaking research with vulnerable young people, there is even less practitioner-researcher research within the area of EAT/L with this group, and no in-depth ethnographies, such as the current study, that I uncovered in my literature review.

I wanted to contribute to the evidence base by providing insights gleaned as a practitioner-researcher through in-depth ethnographic observation of The Yard, together with documentation of the experiences of a number of the young people participating in the TH programme. The young people were understood to be at risk according to the risk and resilience literature due to experiencing various difficulties, challenges, disadvantages and exclusion of some sort. These included having been taken into LA care due to assessed neglect and/or abuse in their parental home, being in contact with a YOT due to criminal activity of some kind, and/or experiencing difficulties in the school environment, with three of the participants having a Statement of SEN. All the participants were considered to require additional therapeutic and/or learning support, with four regularly seeing a psychotherapist or educational psychologist. As many of the young people found it difficult to engage with traditional school and therapeutic environments, they were referred to TH as an alternative and/or additional intervention.

Insights from the ethnographic research study

Through mixed-method data collection involving ethnographic observations, detailed 'thick descriptions', and field and semistructured interviews, together with a number of questionnaires with both the young people and adults involved in their care, a range of data were gathered. The results of the research study revealed that a number of themes arose that were related to the young people's experiences with the horses, together with the orientation and setting of The Yard. These themes have links to some of the claims and theoretical frameworks outlined previously in the AAT and EAT/L literature. As a number of the themes overlapped, they were grouped in relation to their theoretical orientation. The themes were loosely based around issues of nurture, attachment and trust; social well-being and resilience; identification with the horse; a 'safe' space and calming influence; the role of the horse in the therapeutic arena; and the natural environment and spiritual dimensions. Relevant theoretical frameworks adopted were from the risk and resilience literature, attachment theory, psychotherapeutic literature, mindfulness, and the nature/outdoor and ecotherapy fields.

It was found that in all cases the young people developed relationships with the horses on some level. These were sometimes strong attachments, such as in the case of Cinderella and Leo, and in others seemed to involve the young people identifying with different horses at certain times of their lives. Minimax talked about how he believed that the mare Ruby 'understood him', and both Cinderella and Emma spoke about how they could talk to the horses and 'tell them their secrets'. Trusting and having an attachment to a non-judgemental, warm, sensitive, living, breathing horse could perhaps offer experiences similar to those suggested as important in attachment theory in the AAT and EAT/L literature, of satisfying needs for healthy and safe physical, emotional and social contact (Bachi, 2013; Beck & Madresh, 2008; Chardonness, 2009; Walsh, 2009). Establishing these relationships and, in turn, beginning to trust and feel safe with the horses seemed to be a prerequisite to the young people feeling able to try more risky activities with the horses, such as round-pen exercises and riding. These led to gains in self-confidence and self-esteem, both from successfully overcoming challenges and from becoming more proficient in horse management and handling. It is argued that self-esteem is crucial to resilience and that 'mastery' experiences, through which young people feel accomplishment from being successful at challenging tasks, can be useful in promoting resilience as this can generate self-efficacy (Frame, 2006; Masten, Best & Garmezy, 1990; Rutter, 1985). A second area considered important for healthy development and successful life outcomes is the development of empathy, and of emotion regulation, awareness and coping strategies (Goleman, 1996; Lexman & Reeves, 2009; Southam-Gerow & Kendall, 2002). By learning about horse psychology and the different personalities, characteristics and histories of the horses, many of the young people appeared to be able to identify and empathise with them. Wayne empathised with a young colt having left behind his herd and dam in Spain, and with another horse returning to a rescue charity. He wanted to spend whole sessions just being with the horses and discussing their lives, seeming to gain reassurance from processing their experiences. Through her affection for the old pony Leo, Cinderella demonstrated her ability to empathise with his illness and change her usual pattern of behaviour to overcome her fear of medical intervention, remaining to help reassure Leo whilst he received treatment from the vet. Some of their carers also commented that a number of the young people were able to regulate and control their behaviour, both through being motivated to be with the horses, and because they understood that they needed to modify their behaviour in order to have successful relationships with them. An example of this was given by Emma's adoptive mother who said that it was only by Emma being motivated by horses that she would do any academic work in school, which had to centre on horse subjects. Lucy learnt that she could monitor her emotions through the horses, speaking about how she would calm down around the horses as she understood that they did not respond well to angry behaviour. In addition she said that they had taught her how to act confidently around them, which in turn helped her to feel more confident. The therapist Deana believed that this was a two-way process, and that the horses also monitored and changed their behaviour depending on the young person with them. She gave the example of the mare Ruby which she said would act like a different horse with different people, being gentle and quiet with nervous and unconfident children, but sometimes challenging and mischievous with

Conclusion 193

young people who were more outgoing. The ability to regulate emotion and possess emotional understanding is argued to be vital to successful childhood development, Southam-Gerow and Kendall claiming that 'delayed or limited emotion understanding may place youth at-risk for disorder' (2002: 201).

A third prominent theme concerned the horse and the therapeutic relationship. Some authors claim that horses can provide a bridge to building a relationship between client and therapist as the horse provides many of the qualities that are essential in the successful therapist–client relationship (Brooks, 2006; Chardonnes, 2009; Karol, 2007; Yorke, Adams & Coady, 2008). This was corroborated by the EAT/L therapists in the current study. Deana recounted that Kelly called her by her name for the first time in a TH session and posited how horses do not have any preconceptions of the child because 'the horse hasn't read the referral form [...] doesn't have any idea of attachment disorder'. In her study with 'at-risk adolescents', Frame (2006) reports a similar finding of how the experiential nature of EFP and contact with the horses appeared to facilitate rapport-building between the participants and therapists. She writes that the 'interactive experiential component of EFP provided this client unique opportunities to establish rapport with her therapist that overcame her previous discomfort and led her to endorse the importance of talking' (Frame, 2006: 116). The participant Lucy said that the fact that the horses 'just take you for who you are' was something that was important to her. This concept of unconditional positive regard was expressed by the young people in different ways and seemed to be one of the factors that enabled them to build relationships with the horses which then led to other psychotherapeutic benefits. Deana, The Yard's psychotherapist, suggested that the horses provided a 'holding' environment on some level, both literally when participants were sitting or lying on top of a horse, and metaphorically, in that the young people felt safe because they trusted them. Some of the related themes centred on the young people being able to have healthy tactile experiences and express love and affection to the horses, in ways that they may not have been able to with the humans in their lives.

The fourth area that was perceived as having importance to TH by the adults who contributed to the study was of both the physical and the philosophical setting in which The Yard was located. It was found that the natural environment may have incorporated elements of the 'backdrop to activities' suggested by Wals (1994), which encouraged a restorative and calming atmosphere. Being outside in the elements,

194 *Equine-Assisted Therapy and Learning with At-Risk Young People*

with space and freedom, was suggested by the EAT/L practitioner and counsellor Sally as having a big impact on the behaviour of one of the young people with whom she worked. Sally said that this particular young person was much calmer once he was at the TH centre than in the residential facility where he lived, where he would often appear anxious and frustrated. Both Emma and Lucy talked about how being outside in the countryside made them feel different. For Lucy, being in nature facilitated listening to her instincts more than being in the city; for Emma it was about how it offered more freedom than her school environment. In addition, Cinderella said that being in nature was 'destressing'. Some of the adults noted that many of the young people were more responsive and open to learning in the natural environment and context of The Yard. The staff member from Wayne's PRU school informed The Yard that Wayne had written the first essay he had ever attempted since attending the PRU about the horse Jason. Learning in context seemed to be the motivational factor for Freya, who expressed interest in being able to pick and eat wild garlic from the woods adjoining The Yard. Wals (1994) suggests that this is because 'nature piques the curiosity' of young people and so inspires learning. The therapists Deana and Sally believed that a combination of factors was responsible for creating the calming atmosphere of The Yard that in turn facilitated a positive learning and therapeutic environment. These included the peaceful natural environment in which The Yard was located, the use of the natural horsemanship approach to the management of the horses adopted, and the encouragement of a respect for the horses' feelings. The informal, non-directive and experiential learning environment, which contrasted strongly with the characteristics of the school environments in which many of the young people struggled, was seen as highly influential. Elements of these factors were also mentioned by the participant Emma and the adults Linda and Peter.

Many of the insights gleaned from the research study confirmed other EAT/L literature in terms of familiar themes of the role of the horse in the therapeutic process. These include links to attachment theory (Bachi, 2013; Bowers & MacDonald, 2001; Chardonnes, 2009), increases in confidence and self-esteem, development of empathy and the ability to overcome behavioural problems (Bizub, Joy & Davidson, 2003; Burgon, 2003; Cawley, Cawley & Retter, 1994; Kaiser *et al.*, 2004; Klontz *et al.*, 2007; Vidrine, Owen-Smith & Faulkner, 2002). However, the theme of mindfulness in relation to EAT/L has received less attention and connections are limited to a handful of references by authors and practitioners who have noticed links (Broersma, 2007; Kimball,

2002; Soren, 2001). These authors suggest that being with horses can bring people closer to what is described as mindfulness, a way of bringing attention to the present moment. Interacting with horses demands one to be in the present moment due to their large size and particular characteristics – being inherently fearful and requiring clear leadership. Horses also require relaxed concentration, which is different from intense concentration and requires more body awareness, which is also linked to mindfulness practices such as MBSR. Because they are prey animals, horses live perpetually in the 'here and now' as they have evolved to react instantaneously to signs of attack at any moment. Their acute awareness of tiny occurrences draws the attention of participants working with them to the present moment, enhancing mindfulness. Their large size also impresses participants, increasing attention, whilst their need for clear leadership demands being in the present moment. Dealing with horses, then, requires awareness spread over a range of scales: the environment in general and its changing stimuli, the horse and its movements, attention to where one is going and body awareness, all practised within a framework of relaxed concentration.

It was clear that themes linked to mindfulness, such as calmness, were linked to other themes, such as the natural environment and the role of the horse in the therapeutic relationship. In fact, it became apparent that all of the themes overlapped in some way and all were interwoven to a considerable extent. However, what did appear fundamental to the whole process was the horses' particular and unique characteristics, which seemed to provide a motivational backdrop, or 'glue', to the overall experience of TH.

Limitations of the study

There were a number of limitations to this study. In terms of the participants it is acknowledged that this was not a random selection. They were all recruited from The Yard, a TH programme which I, the researcher, had established. Therefore they were all known to me; I had a prior relationship with them to some extent, and could choose who to ask to participate in the research and who to leave out. In fact the reality of the recruitment process was that, due to the limited number of participants attending The Yard over the period of the research project and their transitory and disjointed lives, together with the convoluted consent process, the decision was made to ask initial permission from the majority of the young people who were attending during the start of the research study, which amounted to 15 attendees. Unfortunately,

196 Equine-Assisted Therapy and Learning with At-Risk Young People

however, because many of the young people did not remain at The Yard for the length of time that it took to complete the consent process, they were unable to be included. One social worker refused to give consent due to her fear that it was adding yet another layer of bureaucracy and pressure on a particular young person. In another case I was given verbal consent from the young person involved and got all of the other gatekeeper consents in place but he then moved foster placement to another part of the country and I was unable to gain any further contact with him. This resulted in only seven young people ultimately participating in the research study.

Whilst the main objective of the study was to provide a descriptive exploration of the participants' experiences of TH, together with my reflexive observations, the small sample size makes generalisation of the findings limited. Having a larger number of participants over a variety of different research sites might have provided further observations and insights into TH and enhanced the study. Nevertheless, I make reference to Payne and Williams who claim that 'qualitative research methods can produce an intermediate type of limited generalisation, "*moderatum* generalizations" ' (2005: 296). They suggest that these modest, pragmatic generalisations drawn from personal experience provide useful contributions to how social interaction and everyday life are constructed (Payne & Williams, 2005). It is also acknowledged that it was not possible to separate out any other variables that may have impacted on the participants' experiences at The Yard; it is not possible to know from this study whether participating in TH at The Yard had an independent effect on the young people, or whether a combination of other factors present in their lives at the time were also involved. However, the qualitative, phenomenological approach that I adopted is more concerned with the abstract concepts often found in social work research, such as people's experiences, emotions and feelings, and maintains that these hold as much validity as other findings and methods (D'Cruz & Jones, 2004; Denzin, 2002). Hollway and Jefferson go so far as to suggest that there is no place for attempting to separate variables and seek measurements in human behaviour or feelings, claiming that this results in limited understandings and no context to people's meanings (2000). Moving onto questions of an ontological nature, Gubrium and Holstein (1997) suggest that what is important is not questions concerned with reality but how people's experiences are represented and interpreted, whilst rejecting extreme postmodern claims that nothing exists beyond interpretation. In this study my epistemological position rests in a middle ground between understanding that there is a reality 'out there' and acknowledging that

Conclusion 197

my interpretation of this hinges on a multitude of other factors related to my life experiences.

Problems of respondent bias are another area of consideration. As a practitioner-researcher conducting 'practice-near' research, my position could be argued to compromise relationships and the process of both the research and the practice of TH. There is a multitude of evidence pointing to researcher bias and the fact that 'people respond differently depending on how they perceive the person asking the question' (Denscombe, 2003: 169). There are also questions of researcher contamination – of how people may act differently when under observation (Atkinson, Coffey & Delamont, 2003). However, as this was a reflexive undertaking I agree with May (2001), who argues that it is simplistic to imagine that it is possible to conduct any qualitative research as though it is 'uncontaminated' by the researcher's presence. My focus has revolved more around my observations and reflections on the processes unfolding as I understand them, together with a psychosocial approach to analysing the social interactions which took place at The Yard (Clarke & Hoggett, 2009; Gubrium & Holstein, 1997). In addition, I align myself with feminist-informed discourses which argue that reflexive practice is necessary in order to challenge the power imbalances inherent in current social work relationships (Butler, Ford & Tregaskis, 2007).

Implications for social work practice and related fields

From a clinical and practice perspective, it is hoped that this book offers a useful contribution in terms of providing an in-depth exploration and discussion of the experiences of a number of young people participating in therapeutic horsemanship. An awareness of participants' experiences is important for practitioners and therapists in order to tailor treatment interventions in their own TH and EAT/L practices. In-depth exploration of the philosophy, style and orientation of the practice of The Yard also offers useful information for practitioners in terms of providing concrete examples of how different modalities are applied in practice. Bloor suggests that ethnographies of therapeutic practices can be useful for other practitioners and settings:

> Rich description of particular kinds of therapeutic practice, for example, can assist practitioners in making evaluative judgements about their own practices, preserving what seems to them good practice and experimenting with the adoption of new practices where this seems appropriate.
>
> (1997: 229)

198 *Equine-Assisted Therapy and Learning with At-Risk Young People*

The Yard employed an eclectic mix of therapeutic and learning styles, ranging from experiential, non-directive approaches to more directive, gestalt and mindfulness-influenced exercises, all within a person-centred framework. There are a multitude of approaches being adopted in the burgeoning EAT/L field, including CBT and coaching models, some of which include ridden work while others are completely non-mounted interventions. This research study identifies the particular emphasis of The Yard and will, it is hoped, have useful application to other styles and approaches, as similarities and differences can be drawn to beneficial effect. By providing insights into the practice and activities undertaken at The Yard, together with the particular theoretical therapeutic orientations, the study gives specific and concrete examples of some of the theoretical frameworks as they relate to practice.

One finding of overriding importance from a practice perspective is that horses would seem to offer an opportunity for excluded, disaffected and often resistant young people to engage with professional therapeutic and educational practitioners by allowing them to build up relationships with them. It has been suggested by a number of authors that traditional 'talk' therapies may not always be the most appropriate for some disengaged young people who may find this method too intense (Biegel *et al.*, 2009; Brooks, 2006; Ewing *et al.*, 2007). This research study found examples of how participants were able to engage with therapists and practitioners where they had been unable to do so previously, such as Kelly calling Deana by her name for the first time during a TH session. Frame also found this theme in her study with at-risk adolescents, writing: 'It was posited that the dynamic interchange between the adolescents and horses provided a depth to the clinical work, which exceeded the traditional confines of an office' (2006: 133). How the young people developed positive relationships with both the horses and practitioners also has relevance to relationship-based social work theories. Howe makes reference to this, arguing that the quality of the social work relationship is a fundamental aspect of effective practice, and stating that the 'relationship offers the chance to feel safe and be understood' (2009: 158). As has been seen through the ethnographic fieldwork and interviews, safety was a major theme and was often expressed as the horse's need for safety, which could be understood to be a reflection of how prominent different aspects of safety were for the participants. It has been shown that as the young people began to feel safer with the horses, they were in turn then able to open up to the practitioners and therapists, demonstrating how important it is within the social work relationship for young people to feel safe before they can trust and open up. Feeling understood by the horses was another area found to be important for the young people, as many participants spoke of how they felt the horses understood them. With reference to relationship-based social work, Howe writes that it 'is human to want to understand and be understood by others' (2009: 158). This was expressed by Freya speaking about Ruby when she exclaimed 'It's like she can read my mind', and provides an example of how important it is for social workers and other practitioners to be able to convey 'Empathy – to see and understand how the world looks and feels from the client's point of view' (Howe, 2009: 166).

Another area of importance to social work and psychotherapeutic practice that horses were shown to provide to many of the participants is around the theme of touch. As discussed in Chapter 4, the area of touch in the social work and therapeutic relationship is fraught with difficulty and complexity (Lynch & Garrett, 2010; Pemberton, 2010), yet it is also acknowledged as being a fundamental psychological need (Levinson, 1980; Scott, 1980). Julia Twigg, writing about how the body is positioned within social policy in health and community care, makes the point that 'the body has been evaded for reason of "good practice"' (2002: 425). In their paper on 'touch practices' within social work, Lynch and Garrett (2010) argue that social workers generally agree that reassuring and comforting touch in the area of child protection is important and valuable in building the therapeutic relationship and to demonstrate empathy. In this research study it was found how many of the young people, and particularly the boys, seemed to find comfort from tactile experiences with the horses, such as Minimax and Wayne often spending time hugging the mare Ruby. This perhaps provides some useful information to social work and other related fields about the importance of touch for many of these young people who have experienced dysfunctional and confusing tactile relationships, and about looking for alternative ways of facilitating this, perhaps through other pets and animals if EAT/L is not available/appropriate. The field of AAT makes much reference to the healing element of tactile experiences with animals for both young people and adults, with the much cited study by Friedmann *et al.* (1983) on how stroking a dog can reduce blood pressure being used as a prime example, amongst other claims of the therapeutic benefit of touch with animals (Brooks, 2006; Melson, 2001).

Related to themes concerned with the body, it was found that a number of the young people were able to modify their behaviour by gaining awareness of the role of their body in interactions with horses. This is an area which would seem to have links to mindfulness-based

200 *Equine-Assisted Therapy and Learning with At-Risk Young People*

therapies which are broadly concerned with encouraging mind-body awareness alongside relaxation, and was a new theme to emerge from my research. An example of this is given in Chapter 6, where Wayne learnt how monitoring his breathing and body language reassured and calmed Ruby when she became frightened by the wind. In terms of implications for practice, this perhaps provides support in considering other body-centred alternative therapeutic activities for engaging young people in order to encourage positive behavioural change, such as mindfulness-based stress-reduction programmes (Biegel *et al.*, 2009) and other gestalt-based and outdoor activities which stress the value of an acknowledgement of the mind–body link (Carns, Carns & Holland, 2001; Ewert, McCormick & Voight, 2001; Roszak, Gomes & Kanner, 1995).

Reflections on resilience as related to the study and wider sociopolitical influences

The insights related to resilience frameworks, attachment and psychotherapeutic perspectives provide examples of how these areas can be related to practice with young people through experiences with horses. By beginning to engage in a healthy and motivating intervention, young people are able to acquire benefits that may lead to greater resilience. Increases in self-esteem and self-confidence were reported in the study, together with other psychosocial benefits, such as behaviour modification, the acquisition of empathy, improved relationships and social skills, and the opening up of 'positive opportunities'. These are all of relevance to social work and the therapeutic field in terms of empowerment and enabling positive change for individuals (D'Cruz & Jones, 2004; Denzin, 2002; Fook, 2002; Walker, 2003). Despite there being only limited literature describing it, there is in fact a long tradition of involving young people in outdoor activities with the aim of engagement and bringing about positive change, and the insights gleaned from the research study would appear to support the aims of these interventions (Moote & Wodarski, 1997; Ungar, Dumond & Mcdonald, 2005). Within the current 'Big Society' political and financial climate of promoting community-led initiatives to provide solutions to social issues, it may be that interventions such as TH and EAT/L together with other alternative therapeutic and educational programmes can offer some of the elements suggested in this directive. Many of the current TH and EAT/L centres are charitable and voluntary organisations drawing on community resources and involving a range of people with different skills and experiences, all centred on an equine environment. These factors can possibly make these TH, EAT/L and other alternative programmes more accessible and attractive to young people who would normally be resistant to professional agencies and traditional therapeutic and educational interventions.

By completing this study my perception of resilience has evolved from a purely academic and theoretical understanding to a deeper, embodied perspective. By this I mean that by being deeply engaged and in relationships with the young people through the practitioner-researcher process, I was able to witness first hand how their experiences with the horses seemed to provide them with resilience-related benefits. However, I am also aware of how wider structural and sociopolitical elements also have a major influence and are intrinsically intertwined. This was demonstrated by how fragile and disjointed the lives of some of the young people in foster care were, such as when Kelly discovered that she was moving foster placement the following day whilst in a TH session. For some of the young people, gains that they may have made, such as feelings of increased self-efficacy and being able to have an influence on their future, together with enhanced social relationships and trust, may have been shattered by experiences such as this. Another example of how socioeconomic and political factors impinged on The Yard, and therefore on the lives, progress and impact on the resilience of the young people attending, is provided by the political structure within which The Yard was located. Being funded by a private foster care agency, The Yard became a victim of the profit-making drive behind this privatisation structure, when the company was sold to a larger conglomerate which decided for financial or theoretical reasons that TH was not necessary within the organisation and ceased to fund it, thus resulting in the closure of The Yard at the time. This relates to Bronfenbrenner's (1979) ecological systems theory that argues that it is impossible to separate the micro from the macro in terms of how all societal systems are interrelated. Having a humanistic (and largely positive) perspective, I could only hope that any small gains made in personal resilience factors would provide longer-term benefits to the young people as proposed in a 'dynamic state' model (Cicchetti & Rogosch, 1997; Place *et al.*, 2002).

Methodological and theoretical contributions together with implications for policy

A major methodological tool that I employed was writing detailed reflexive fieldnotes after each TH session. These evolved from being rather clinical and based on recording the 'facts' of the TH session into long and highly descriptive accounts exploring as many elements as I could manage to record.

However, the full value of these detailed fieldnotes only became wholly apparent when I embarked upon the data analysis and when I began to see elements and patterns which would easily have been overlooked without my descriptive and, at the time, seemingly trivial records of events. Examples of this are where I recorded how some of the young people mentioned the smell of the horses, and instances where we noticed wildlife on the farm. Initially these were not things that I considered important to record, concentrating on the 'concrete' interactions between the horses and participants. However, as I became more immersed in the practitioner-researcher role, my fieldnotes grew correspondingly and, in retrospect, they have provided a crucial part in the ethnographic process. It is hoped that this knowledge, of the value of detailed, descriptive and reflexive fieldnotes, will prove useful for other ethnographers in the field of social work and other related research.

Another area in which it is hoped this book will be proved useful is in its impact on and contribution to participatory research with young people in a social work context. An intention of the research project was for the young people who participated to gain some benefit by way of their involvement being an empowering experience, as opposed to the disempowering experiences that they often encountered within their various agencies. For example, Wayne and Minimax were initially difficult to engage in the research process when the research was explained to them – they simply seemed to 'switch off' when forms and paperwork were produced. However, when the research was applied to a real situation, such as asking for their input to the questionnaire and rephrasing the research in terms of their contribution as providing an understanding of how horses could possibly help other young people, they were immediately motivated, and keener to engage and offer suggestions. In line with participatory values, there has been greater emphasis in social work and allied fields on service user research in the past decade, with the government having stated that it wants to increase the involvement of young people in the evaluation of services and policy (McLaughlin, 2005). Despite this notion that participatory research is 'better', and the call for more research with young people to adopt 'participatory' methods in recent years, there is criticism of the levels and reality of participation and its assumed superiority (Holland *et al.*, 2010). Notwithstanding these criticisms, Walker argues that it is necessary in social work for more research on service provision to include the views of young people, not least because services for children and young people's mental health have been based on adult concepts, models and practices. This has led to paternalistic, patronising practices that seek to emasculate children and stifle their creativity and wisdom (2003: 136).

Walker claims, therefore, that the input and views of young people in the services that they are engaged in is imperative in order to ensure that service provision is effective, and in order to find out 'what works best for which children in what circumstances' (2003: 136). Hopefully this research study has provided some insight into the process and the positive and achievable aspects, as well as the pitfalls and more difficult areas of participative research with disenfranchised young people.

On a personal level, undertaking this study has provided me with increased appreciation and understanding of the potential and depth of the role of the horse in the therapeutic and learning arena. Through the academic endeavour of data analysis and writing, I was able to keep a useful distance from the practice together with the opportunity of in-depth analysis of the intricacies that developed between the young people and the horses. This has led to an increased awareness of how sometimes allowing the work to unfold instead of being too goal-orientated could often result in a very powerful experience for participants. It has also made me realise how much more there is to learn, and that there are areas of the horse–human relationship which will perhaps always remain unknowable. Nevertheless, it has confirmed my commitment and passion for this work and has enabled me to return to the practice of TH with increased knowledge and understanding of horse–human dynamics. By having an enhanced understanding of the theoretical basis of TH, I am hopeful that this academic experience has benefited my capability as a practitioner and will, in turn, contribute to the wider field of EAT/L as a whole.

Finally, on a wider level it is hoped that by contributing to the research base this study will go some way towards supporting further research that will begin to provide a robust evidence base for the effectiveness of TH and EAT/L, as is called for within the field (Bachi, 2012, Selby & Smith-Osborne, 2013). In turn, it is hoped that such a base would help to influence policy decisions regarding the provision of further funding

Counter-indications of TH and EAT/L

It is important to acknowledge that horses do not suit everyone. Some young people may have health conditions, such as allergies, which render TH and EAT/L inappropriate activities. Although it was not an issue in this research study, some young people may just not like horses. Cultural aspects of EAT/L and AAT are also important to note. Many of the theoretical frameworks that support the interventions, such as attachment theory, are argued to be 'laden with Western values and meanings' (Rothbaum *et al.*, 2000: 1093). Other authors mention how some cultures view certain animals as 'unclean' (Friesen, 2010), and an EAL study with North American Indian 'First Nation' youth found it necessary to adapt the Western understanding of resilience as an individual construct to include 'their inner spirit (internal) and relations with their collective community' (Dell *et al.*, 2008: 82).

Another area, often neglected in the literature, concerns the ethical considerations and treatment of the horses involved in practicing EAT/L, and of the effect that participation may have on the animals (Gehrke, Baldwin & Schiltz, 2011; O'Rourke, 2004). There is much research centred on the abuse and cruelty of animals by young people and adults who have suffered abuse themselves (Merz-Perez & Heide, 2004; Risely-Curtiss, 2010), so it is vital that the horse's welfare is paramount and supervision of the young people is adequate. Whilst there are different theoretical and practice models involved in EAT/L, in my opinion it is important for all of the practitioners involved to have in-depth equine experience, including knowledge of the horses in the particular practice, in order to provide a safe environment for both participants and horses.

A practical consideration and large obstacle to the provision and availability of EAT/L is its not inconsiderable cost. The staffing costs of professional and qualified practitioners, together with the upkeep of the horses, premises, public liability insurance and so on, mean that it has a considerably higher hourly cost than traditional talk therapy. Nevertheless, if all of the many different aspects to TH and EAT/L which have been explored through the research study are taken into account, such as their offering an alternative option for some young people who may find it difficult to engage with traditional educational and therapeutic approaches, together with the suggested benefits gained from being in the natural environment, the psychosocial benefits from contact and relationships with animals, the physical and mental health benefits of exercise, and the challenging and risk-taking elements, then it may seem to be an economical intervention after all.

Suggestions for future research

As a marginal intervention with such a limited research base, any further good-quality research on EAT/L is to be welcomed at this stage. A criticism levied at current research is that studies are of mixed quality with small sample sizes, a lack of control groups, and varying populations within studies, resulting in difficulty in standardisation (Bachi, 2012; Lentini & Knox, 2009; Smith-Osborne & Selby, 2010). A large proportion of the literature comprises anecdotal and descriptive accounts by practitioners, together with 'grey' literature, such as dissertations and theses (Esbjorn, 2006; Frame, 2006; Graham, 2007; Hayden, 2005; Held, 2006). Despite the limited amount of good-quality qualitative research on EAT/L, the fact that these studies find similar themes of psychosocial benefits to those uncovered in the current study surely warrants further investigation of the field. One of the advantages of qualitative research is that it can provide in-depth exploration of the processes that are occurring in the social environment and this can then provide a framework for future studies. Whilst the difficulties of conducting quantitative, empirical research within the area of psychosocial studies has been documented previously, it would nevertheless be useful to have a wider contribution of studies with larger populations and more standardised measurement techniques. In addition, research is currently hampered by a lack of regulation and standardisation within the field of EAT/L and a range of perspectives in terms of theoretical underpinnings. There are currently no governing or accrediting bodies overseeing the practice and training of EAT/L practitioners and trainers, so the field is confusing and unregulated, although a number of individual training organisations, including PATH Intl and CBEIP, are working to address this. More standardisation and centralisation of EAT/L will facilitate easier access and understanding for researchers.

In terms of particular themes that arose from the study, the area of empathy within the different aspects of EAT/L practice has received limited attention within the literature. In relation to social work research it is claimed that service users consistently value social workers 'who appear to understand them and who care' as a key factor in the social worker and client relationship (Forrester et al., 2008: 42). Through the young people claiming that the horses understood them and that they could 'tell them their secrets', in addition to appearing to show an understanding of the horses' feelings themselves, different elements of empathy seemed to be at play, so this may be an interesting area for further research. Another relatively unresearched area of particular interest and relevance to social work to emerge from the study is that of community participation and social inclusion. It appeared that the young people enjoyed attending The Yard and were motivated to participate in the activities offered. For some of them, participating in TH seemed to enable them to make social gains such as better peer relations, and improved interactions with the therapists and practitioners. In addition, their interest and motivation in being with horses led to a number of the young people going on to participate in other horse- and animal-related experiences in the wider community. During her time of attending The Yard, Emma started helping at a local RDA centre as part of her school curriculum. Lucy went on to successfully complete a small animal care course at a local college. We were informed that another young person had been accepted on an equine course. Social work is underpinned by values of the importance of increasing community participation and the inclusion of marginalised and excluded groups of vulnerable people into mainstream society (BASW, 2002). Similarly, The World Health Organisation states that community participation is an important part of health and functioning (WHO, 2001). Research that paid attention to the longer-term outcomes of attending TH and EAT/L through follow-up studies with participants would provide useful information and guidance for further practice and policy. Last, but not least, more research on the impact and effect of the different styles and practices of EAT/L on the horses that participate would be welcomed, as this is an area which has received less attention in the literature. Clearly, in order to be able to provide an effective, safe and healing environment it is vitally important that the horses employed are physically and emotionally healthy and content themselves. This is something that Cinderella referred to when asked what she most liked about The Yard, replying:

> The horses, also there is no nastiness at The Yard, everyone is polite and the horses just get on with each other.
>
> (Cinderella)

Appendix A
Research Project into Equine-Assisted Therapy/Learning and Therapeutic Horsemanship

Information sheet for children and young people

Would you consider taking take part in some research?

I am doing some research into equine-assisted therapy/learning (EAT/L) and therapeutic horsemanship (TH) as part of a PhD at Cardiff University and would like to invite you to take part in the study.

What are EAT/L and TH?

EAT/L and TH are activities with horses which involve learning how to care for horses, and also learning about how they communicate and behave with each other and with us. Horses communicate through body language and will mirror behaviour, so by learning how to communicate with a horse we can also learn more about ourselves.

Activities include grooming, learning how to work with and look after horses, feeding and stable management tasks. Sometimes riding and educational activities are included.

Why am I doing the research?

The reason I am doing the study is that there is very little research into EAT/L and TH and so it is hoped that this research project will help towards supporting more studies. Hopefully this will lead to more young people and adults being able to attend EAT/L and TH in the future.

Who is being invited to participate and what will it involve?

Children and young people who are participating in EAT/L and TH programmes and/or have other contact with horses are being invited to take part in the study. The research will have no effect on the sessions, as the reason for the study is to explore what is naturally happening between the horses and participants.

I will be a practitioner in the sessions and may ask you if you would be willing to answer some questions about the experience, or perhaps complete a small evaluation questionnaire.

208   *Appendix A*

If you do consider taking part I would be more than happy to come and talk to you about it first with your parent/foster carer and/or social worker/teacher present so that you can have the opportunity to ask me any questions you may have.

There is no payment involved on either side in taking part in this study. However, by participating you may be helping to gain more understanding of EAT/L.

What if I don't want to take part?

You are, of course, free to choose not to take part in the study and this will have no effect on whether you can attend EAT/L or TH sessions now or in the future. If you do agree, you can choose how long you wish to take part, from just one session or for the length of your attendance in the project/up to the length of the study. You can choose to stop being in the study at any time and this will have no effect on your participation at the stables you attend.

Confidentiality

When I write up the study, everyone's names will be changed (including the horses) so that no one can be identified. You will be welcome to choose your own name for the study and I am very keen that you will feel able to join in with offering any suggestions about the research. All of the paperwork will be kept in a locked filing cabinet and I will ask if you are happy with what I have written. In the future the study may be published in some publications and journals.

Contact information

If you would like more information about the research or would be willing to have a chat with me about it before deciding, please contact me.

Hannah Burgon
XXX XXX
Email: hannahburgon@hotmail.com or XXX

Appendix B
Consent Form – Young People

Cardiff School of Social Sciences
Cardiff University
Glamorgan Building
King Edward VII Avenue
Cardiff CF10 3WT

Consent Form (Young People)

Equine-Assisted Therapy/Learning and Therapeutic Horsemanship Research Project

Name of researcher: Hannah Burgon

I confirm that I have read and understood the information sheets enclosed and had any questions about the research answered to my satisfaction.

I give my consent to take part in the study and understand that I can withdraw from the research/interview at any time.

I give my consent for information obtained in the research/interview/ questionnaire to be published in academic journals/other publications in the future (all identifying features/names made anonymous).

I understand that I will receive no payment for agreeing to participate in this research.

....Age.....................

Name of participant (print name)

....Date.....................

Signature of above

Person with authority to give consent (social worker/parent/carer)

....Date.....................

210   *Appendix B*

Person with authority to give consent (print name)

. . . . .Date. . . . .

Signature of above

Copy for participant and copy for research file

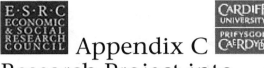

Appendix C
Research Project into Equine-Assisted Therapy/Learning and Therapeutic Horsemanship

Information sheet for participants (adults)

Equine assisted therapy/learning (EAT/L) and therapeutic horsemanship (TH) are interventions aimed at assisting children, young people and adults with additional support needs to gain physical and mental health benefits from being outdoors in nature learning about and working with horses. This type of experiential learning and therapy is well established in the USA and is beginning to be employed in the UK.

However, published research remains scarce and so the purpose of this study is to gain a deeper insight into the experiences and views of practitioners, carers, parents, social workers and other people involved in some way with EAT/L and TH.

A separate study is taking place exploring the experiences and perceptions of participants in EAT/L and TH.

As part of this study you will be asked to complete a short questionnaire and/or take part in an interview with the researcher, Hannah Burgon. This will last about 45 minutes and take place at a time and place convenient to you.

All information that is collected about you during the course of the study will be kept strictly confidential and anonymous. You will not be identified in any report or publication arising from the research.

There is no payment for taking part in this research but hopefully it will help to further your understanding of EAT/L and TH. In addition the information from the study will contribute to the wider research base into the interventions.

The project has the full support and consent of the equine projects involved and Cardiff University's School of Social Sciences Ethics Committee.

Contact information

If you would like more information about the research project, please contact:

Hannah Burgon
XXXXXX email: hannahburgon@hotmail.com or XXX
Cardiff University
Cardiff School of Social Sciences
Glamorgan Building
King Edward VII Avenue
Cardiff CF10

Appendix D
Consent Form – Adults

Cardiff School of Social Sciences
Cardiff University
Glamorgan Building
King Edward VII Avenue
Cardiff CF10 3WT

Consent form (Adult)

Equine Assisted Therapy/Learning and Therapeutic Horsemanship Research Project

Name of researcher: Hannah Burgon

I confirm that I have read and understood the information sheets enclosed and had any questions about the research answered to my satisfaction.

I give my consent to take part in the study and understand that I can withdraw from the research/interview at any time.

I give my consent for information obtained in the research/interview/questionnaire to be published in academic journals/other publications in the future (all identifying features/names made anonymous).

I understand that I will receive no payment for agreeing to participate in this research.

....

Name of participant (print name)

....Date............

Signature of above

Copy for participant and copy for research file

Appendix E: XXXX Ltd Therapeutic Horsemanship Programme

Child protection and disclosure policy and procedure for therapeutic horsemanship

1. A full risk assessment is undertaken prior to a TH referral. This will include any concerns around relevant contact restrictions/child protection/health/behavioural/other issues.
2. The child will be informed by the TH practitioner in a child-friendly and accessible way that if they raise anything of concern (i.e. risk to self or others) during a session then the TH practitioner will be required to take the action outlined in 3.
3. If a child voices anything that raises concern in a TH session, then the TH practitioner will listen and acknowledge that they have heard the child. The TH practitioner will then suggest that the child talks to their carer about the issues after the session has finished. The TH practitioner will inform the child that what they have said is of concern and that their carer/social worker will need to be informed for the child's own safety.
4. Any behaviour by the child that causes the TH practitioner to be concerned about risk to the child/staff/horses/any other animals will result in the session being stopped and the carer being contacted to return to collect the child immediately. If this is not possible the child will be taken to a safe environment until they are collected.
5. At the end of the TH session a debrief with the carer and child can take place to discuss the session. If there is anything that the TH practitioner needs to discuss separately with the carer, the TH horse handler will remain with the child whilst the TH practitioner does this.
6. If the TH practitioner thinks that there are any concerns that need to be relayed immediately, this will be done, either directly to the carer after the session, as above, or, if this is not possible, by telephoning the carer/supervising social worker/child's social worker/XXX director with the child's consent if possible.
7. All sessions are written up and a debriefing is held with the TH practitioner and horse handler after the session. This is then sent to the carer's supervising social worker as soon as possible. The supervising social worker then sends copies to the child's social worker (and carer if required). A copy is also kept on file. Referrals are discussed at monthly team meetings and recommendations

214 *Appendix E*

made to the child's social worker to refer the child for TH sessions with the project play therapist if this is deemed appropriate.

8. Any child-protection issues will be handled in accordance with XXXX standard child protection policy.

Hannah Burgon – Child Protection Policy 2006
(reviewed and updated November 2007)

Appendix F: XXXX Ltd Therapeutic Horsemanship Programme

Horse management policy and code of ethics

1. The horses' welfare is paramount – any concerns regarding their health, physical or otherwise, should take priority over a session. A decision is to be taken in conjunction with the TH practitioner and the horse handler as to what course of action is to be taken. Hannah Burgon is to be consulted at the earliest opportunity but if she is not available and a decision needs to be made to call a vet then this is to be undertaken. XXX is also to be consulted if possible.
2. The horses are to be kept as naturally as possible within the constraints of the environment available – that is, a natural horsemanship management policy is followed with the horses being kept together as part of a herd and in a free-range barn system. Their feeding reflects this and worm counts are regularly undertaken in order to avoid chemical worming as much as possible. Hay unsprayed by chemicals is used as the first choice when it is possible to obtain. Clipping is avoided unless absolutely necessary and no trimming is undertaken (unless for first aid/veterinary reasons or when tails are trailing on the ground).
3. No whips or spurs are to be used (a lunge stick can be used in the round pen to encourage the horse to move forward or change direction only).
4. Tack: hackamores are to be used as standard and saddles should be checked regularly for fitting. The importance of taking care over fitting tack, how it is put on and the effect of the rider's weight on the horse's back are to be carefully explained and demonstrated to participants in sessions.
5. The horse's physical and emotional needs are to be constantly under supervision during a session. This is the responsibility of both practitioners but especially the horse handler. Participants are encouraged and supported to learn about horse behaviour and to be aware of how the horse is feeling, what the horse's body language is telling them and the other horses, and of the effect of their own behaviour and body language on the horse.
6. Last but not least, remember that how we treat and care for the horses reflects on what we are trying to convey in the sessions – that is, principles of respect, compassion, patience and understanding.

Horse code of ethics – Hannah Burgon, May 2006

Appendix G: Questionnaire about XXX Therapeutic Horsemanship (children and young people)

1. Do you have a favourite horse at XXX? If so, which is it and what is it about this horse that you like? If you don't have a favourite horse, could you write something about any of the other horses?
2. What do you like doing best with this horse/any of the other horses and why?
3. Does being around the horses change your mood or ever make you feel differently? If so, what is it about the horses that changes how you feel?
4. Do you think that the horses notice your moods? How do you know? Do you think that our behaviour affects the horses. If so, how?
5. How do you think the horses like us to treat and behave around them?
6. Is there anything you have especially learnt from spending time with the horses?
7. Is there anything else you would like to say?

Name: ....Date:................................

Thank you very much for your time and patience in completing this questionnaire!

Hannah Burgon, 2006

Appendix H: XXX Therapeutic Horsemanship Programme

Questionnaire for carers/social workers/parents

Name of young person...

1. Do you think that the young person has learnt anything from attending therapeutic horsemanship? Are these things that could be applied to other parts of their life (e.g. learning about responsibility, empathy, how our behaviour affects the horse and vice versa, etc).
2. Do you notice any changes/differences in the young person after participating in therapeutic horsemanship sessions (i.e. in behaviour/body language etc.)? If so, what do you think it is about attending therapeutic horsemanship that may have contributed to this?
3. Do you think that it has been beneficial for the young person to attend therapeutic horsemanship? If so, please explain what the benefits are. If not, please state the reasons why not?
4. Do you think that horses/animals are therapeutic in general and if so can you say why they may be particularly so for young people with additional needs?
5. Is there anything else you would like to add? Please continue on another sheet if necessary.

Name...Date.............

Position..

Thank you very much for your time

Appendix I: PhD Questions for Young People

Introduction

I am doing some research about horses and young people and would like to talk to you about your experiences with horses, what sort of things you like and don't like about horses, what you have learnt from horses and so on.

1. Would you mind telling me a little bit about yourself? How would you like me to describe you for the research? Would you like to choose a pseudonym (false name) for yourself?
2. Do you have a favourite horse and if so could you tell me what is it about that horse that you like? How would you describe XXX? What sort of personality have they got?
3. Thinking about all of the horses here, which horse do you think you are most like and why?
4. Which horse would you like to be like and what is it about that horse that you want to be like?
5. Who do you think is the lead horse in the herd – the horse the others trust and will follow? What is it about this horse do you think that the others have chosen them as leader?
6. What do you think about horses? How do you see them? Are they like a work animal, a dumb creature, a friend, a teacher or anything else?
7. How do you think the horses/particular horse chosen see/s you?
8. How do you think the horses like us to behave around them? Do you think that horses have feelings? If so, how can you tell?
9. Do you think horses can tell our moods? If so, how do they tell you? How do you think they can tell if we are stressed or in a bad mood, for example? Have you learnt any different ways of dealing with stress when you are around horses?
10. Do you ever feel different after you have been to the stables and spent time with horses from how you did before you got there? In what way? What was it about the horses/stables that made you feel different?
11. What do you like best about being/working with horses? Are there any special experiences you can think of? What are your favourite things? What has been your best experience with XXX (horse)?
12. If you were telling someone else about what it's like to look after/have a horse, what would you say?
13. Can you tell me what the most important things you have learnt about horses/being with horses are?

218

Appendix I 219

14. Can you think of some of the ways in which being with horses/having a horse/coming to the yard helped you in other parts of your life (e.g. with friends/school/how you feel about yourself)?
15. Is there anything else you would like to tell me about the horses?

Appendix J: Tables of Participants

Young people participating

Name	Age at start	No of sessions	Time attended	Referrer	Research contribution
Cinderella	15	6	1 year	Foster care agency referral	Informal field interviews, questionnaire
Emma	15	20+	1 year	Private/Social Services- voluntary participant	Informal field interviews, semistructured interview
Freya	14	16	1 year	Residential children's home	Informal field interviews
Kelly	14	9	6 months	Residential children's home	Informal field interviews
Lucy	19	30+	2 years	Private – voluntary participant	Informal field interviews, semistructured interview
Minimax	12	15	2 years	Foster care agency	Informal field interviews, questionnaire
Wayne	12	8	6 months	Youth offending/PRU	Informal field interviews

Adults – main participants

Name	Role	Research contribution
Angie	Foster carer - Minimax	Interview and questionnaire
Deana	Therapist at The Yard	Interview and case notes
Linda	Adoptive mother – Emma	Interview
Peter	YOT mentor – Wayne	Interview and questionnaire
Sally	Counsellor at The Yard	Interview and case notes

Adults – supplementary contributions

Carl	Manager – residential home 'The Elms'	Questionnaire
David	Foster carer	Questionnaire
Laura	Foster carer	Questionnaire

Glossary

Equine-assisted learning (EAL) and equine-facilitated learning (EFL) are concerned with experiential, educational and social learning based around horses. They are practised by social workers, mental health practitioners, teachers and other educators.

Equine-assisted psychotherapy (EAP) and equine-facilitated psychotherapy (EFP) employ a qualified mental health professional and are centred on gaining self-awareness within a psychotherapeutic context. They are generally short-term means of intervention and do not usually include riding.

Therapeutic horsemanship (TH) is concerned with all aspects of horse management and learning within a therapeutic context and can include riding.

Equine-assisted therapy (EAT) is where horses are employed for therapeutic benefit under the supervision of qualified practitioners. In the UK it has often been employed as an umbrella term for many of the above interventions.

Hippotherapy, riding therapy and therapeutic riding are all concerned with physical and mental health benefits from riding a horse, whilst acknowledging the social and emotional benefits derived from the activity.

Animal-assisted therapy (AAT) is associated with smaller animals, such as cats and dogs. It includes the organization PAT, which takes dogs into hospitals and residential care homes for therapeutic and social benefits.

Bibliography

Ainsworth, M. and Bell, S. (1970) Attachment, Exploration, and Separation: Illustrated by the Behavior of One-Year-Olds in a Strange Situation. *Journal of Child Development*, 41: 49–67

Ainsworth, M. and Bowlby, J. (1991) An Ethological Approach to Personality Development. *American Psychologist*, 46 (4): 333–341

Alderson, P. (2001) Research by Children. *International Journal of Social Research Methodology*, 4 (2): 139–153

Alderson, P. (2004) Ethics. In S. Fraser, V. Lewis, S. Ding, M. Kellett and C. Robinson (Eds.), *Doing Research with Children and Young People*. London: Sage

All, A., Loving, G. and Crane, L. (1999) Animals, Horseback Riding and Implications for Rehabilitation Therapy. *Journal of Rehabilitation*, 65 (3): 49–87

Almaas, A. (1988) *The Pearl Beyond Price-Integration of Personality into Being: An Object Relations Approach*. Boston, MA: Shambhala Publications

Anderson, R., Hart, B. and Hart, L. (Eds.) (1984) *The Pet Connection: Its Influence on Our Health and Quality of Life*. Minneapolis: University of Minnesota

Anderson, W., Reid, C. and Jennings, G. (1992) Pet Ownership and Risk Factors for Cardiovascular Disease. *Medical Journal of Australia*, 157: 298–301

Arluke, A. (1994) We Build a Better Beagle: Fantastic Creatures in Lab Animal Ads. *Qualitative Sociology*, 17: 123–158

Arluke, A. and Saunders, C. (1996) *Regarding Animals*. Philadelphia: Temple University Press

Armstrong, S. and Botzler, R. (Eds.) (2003) *The Animal Ethics Reader*. London: Routledge

Atkinson, P., Coffey, A. and Delamont, S. (2003) *Key Themes in Qualitative Research: Continuities and Change*. Walnut Creek, CA: AltaMira Press

Atwood, E. (1984) *Rodeo: An Anthropologist Looks at the Wild and the Tame*. Chicago: University of Chicago Press

Axline, V. (1971) *Dibs in Search of Self*. London: Penguin

Bachi, K. (2012) Equine-Facilitated Psychotherapy: The Gap Between Practice and Knowledge. *Society and Animals*, 20: 364–380

Bachi, K. (2013a) Application of Attachment Theory to Equine-Facilitated Psychotherapy *Journal of Contemporary Psychotherapy*, 43 (3): 187–196

Bachi, K. (2013b) Equine-Facilitated Prison-Based Programs Within the Context f Prison-Based Animal Programs: State of the Science Review. *Journal of Offender Rehabilitation*, 2 (1): 46–74

Bachi, K., Terkel, J. and Teichman, M. (2012) Equine-Facilitated Psychotherapy for at-Risk Adolescents: The Influence on Self-Image, Self-Control and Trust. *Clinical Child Psychology and Psychiatry*, 17 (2): 298–312

Baer, R. (Ed.) (2006) *Mindfulness-Based Treatment Approaches: Clinicians Guide to Evidence Base and Applications*. London: Academic Press

Baker, W. (2004) *Healing Power of Horses: Lessons from the Lakota Indians*. Irvine, CA: Bow Tie Press

224 Bibliography

Ball, D., Gill, T. and Spiegal, B. (2008) *Managing Risk in Play Provision: Implementation Guide*. The Department for Children, Schools and Families (DCSF) and the Department for Culture, Media and Sport (DMCS), available online at: http://www.playengland.org.uk/resources/managing-risk-in-play-provision-implementation-guide [accessed on 17 January 2011]

Bandura, A. (1982) Self-Efficacy Mechanism in Human Agency. *American Psychologist*, 37 (2): 122–147

Bannister, A. (Ed.) (1998) *From Hearing to Healing: Working with the Aftermath of Child Sexual Abuse*. (2nd edn.). Chichester: John Wiley & Sons

Barclay, H. (1980) *The Role of the Horse in Man's Culture*. London: JA Allen

Barker, S. and Dawson, K. (1998) The Effects of Animal Assisted Therapy on Anxiety Ratings of Hospitalized Psychiatric Patients. *Psychiatric Services*, 49: 797–801

Barr, R., Boyce, T. and Zeltzer, L. (1996) The Stress-Illness Association in Children: A Perspective from the Biobehavioural Interface. In L. Haggerty, R. Sherrod, L. Garmezy, and M. Rutter (Eds.), *Stress, Risk and Resilience in Children and Adolescents: Processes, Mechanisms, and Interventions*. Cambridge: Cambridge University Press

Bass, M., Duchowny, C. and Llabre, M. (2009) The Effect of Therapeutic Horseback Riding on Social Functioning in Children with Autism. *Journal of Autism and Developmental Disorders*, 39 (9): 1261–1267

BASW (2002) The Code of Ethics for Social Work. *British Association of Social Workers*. Birmingham: BASW, available online at: http://www.basw.co.uk/about/codeofethics/ [accessed on 29 October 2010]

Beck, A. (1976) *Cognitive Therapy and the Emotional Disorders*. New York: International Universities Press

Beck, A. and Katcher, A. (1996) *Between Pets and People: The Importance of Animal Companionship*. West Lafayette, IN: Purdue University Press

Beck, A., Rush, A., Shaw, B. and Emery, G. (1979) *Cognitive Therapy of Depression*. New York: Guildford Press

Beck, L. and Madresh, E. (2008) Romantic Partners & Four-Legged Friends: An Extension of Attachment Theory to Relationships with Pets. *Anthrozoos*, 21 (1): 34–56

Bekoff, M. (2007) *The Emotional Lives of Animals: A Leading Scientist Explores Animal Joy, Sorrow, and Empathy – and Why They Matter*. Novato, CA: New World Library

Bentley, J (2001) *Riding Success Without Stress: Introducing the Alexander Technique*. London: JA Allen

Berger, J. (1986) *Wild Horses of the Great Basin*. Chicago: University of Chicago Press

Berger, R. and McLeod, J. (2006) Incorporating Nature into Therapy: A Framework for Practice. *Journal of Systemic Therapies*, 25 (2): 80–94

Bergin, A. and Garfield, S. (Eds.) (1994) *Handbook of Psychotherapy and Behavior Change* (4th edn.). New York: Wiley

Berman, D. and Davies-Berman, J. (1995) Adventure as Psychotherapy: A Mental Health Perspective. *Journal of Leisurability*, 22 (2), available online at: http://lin.ca/resource-details/2774 [accessed on 23 October 2009]

Bernstein, P., Friedman, E. and Malaspina, A. (2000) Animal-Assisted Therapy Enhances Resident Social Interaction and Initiation in Long-Term Care Facilities. *Anthrozoos*, 13 (4): 213–224

Bertoti, D. (1988) Effect of Therapeutic Horseback Riding on Posture in Children with Cerebral Palsy. Paper Presented at the *6th International Therapeutic Riding Congress*, Toronto, Canada, 23–27 August

Bexson, T. (2008) Horse Sense, *Mental Health Today,* February 2008: 16–17

Biegel, G., Brown, K., Shapiro, S. and Schubert, C. (2009) Mindfulness-Based Stress Reduction for the Treatment of Adolescent Psychiatric Outpatients: A Randomized Clinical Trial. *Journal of Consulting and Clinical Psychology,* 77 (5): 855–866

Birke, L. (1994) The Renaming of the Shrew. In *Feminism, Animals and Science: The Naming of the Shrew.* Buckingham: Open University Press

Birke, L. (2007) Learning to Speak Horse': The Culture of 'Natural Horsemanship. *Society & Animals,* 15: 217–239

Birke, L. (2008) Talking About Horses: Control and Freedom in the World of 'Natural Horsemanship'. *Society & Animals,* 16: 107–126

Birns, B. (1999) Attachment Theory Revisited: Challenging Conceptual and Methodological Sacred Cows. *Feminism & Psychology,* 9 (1): 10–21

Bizub, A., Joy, A. and Davidson, L. (2003) 'Its like being in another world': Demonstrating the Benefits of Therapeutic Horseback Riding for Individuals with Psychiatric Disability. *Psychiatric Rehabilitation Journal,* 26 (4): 377–384

Blair, D. (2009) The Child in the Garden: An Evaluative Review of the Benefits of School Gardening. *Journal of Environmental Education,* 40 (2): 15–38

Blake, H. (1975) *Talking with Horses: A Study of Communication between Man and Horse.* London: Souvenir Press

Blaxter, L., Hughes, C. and Tight, M. (1996) *How to Research.* Buckingham: Open University Press

Bloor, M. (1997) Addressing Social Problems through Qualitative Research. In D. Silverman (Ed.), *Qualitative Research: Theory, Method and Practice.* London: Sage

Boris, N. and Zeanah, C. (1999) Disturbances and Disorders of Attachment in Infancy: An Overview. *Infant Mental Health Journal,* 20 (1): 1–9

Born, M., Chevalier, V. and Humblet, I. (1997) Resilience, Desistance and Delinquent Career of Adolescent Offenders. *Journal of Adolescence,* 20: 679–694

Bourdieu, C. (1999) *The Weight of the World: Social Suffering in Contemporary Society.* Cambridge: Polity Press

Bowers, M. and MacDonald, P. (2001) The Effectiveness of Equine-Facilitated Psychotherapy with At-Risk Adolescents. *Journal of Psychology and Behavioural Sciences,* 15: 62–76

Bowlby, J. (1984) *Attachment and Loss: Volume 1. Attachment* (2nd edn.). Harmondsworth: Penguin Books

Bowlby, J. (1988) *A Secure Base: Clinical Applications of Attachment Theory.* London: Routledge

Bracha, S., Ralston, T. and Matsukawa, J. (2004) Does 'Fight of Flight' Need Updating? *Psychosomatics,* 45: 448–449

Brandt, K. (2004) A Language of their Own: An Interactionist Approach to Human-Horse Communication. *Society and Animals,* 12 (4): 299–316

Braun, V. and Clarke, V. (2006) Using Thematic Analysis in Psychology. *Qualitative Research in Psychology,* 3: 77–101

Bretherton, I. (1985) Attachment Theory: Retrospect and Prospect. *Monographs of the Society for Research in Child Development,* 50: 3–38

Bretherton, I. (1992) The Origins of Attachment Theory: John Bowlby and Mary Ainsworth. *Developmental Psychology,* 28 (5): 759–775

226 Bibliography

British Medical Association (2005) *Preventing Childhood Obesity*. London: BMA

Brock, B. (1988) Therapy on Horseback: Psychomotor and Psychological Change in Physically Disabled Adults. Paper Presented at the *6th International Therapeutic Riding Congress,* Toronto, Canada, 23–27 August

Broersma, P. (2007) *Riding into Your Mythic Life: Transformational Adventures with the Horse.* Novato CA: New World Library

Brooks, S. (2006) Animal-Assisted Psychotherapy and Equine-Assisted Psychotherapy. In N. Webb (Ed.), *Working With Traumatized Youth in Child Welfare.* New York: The Guildford Press

Bronfenbrenner, U. (1979) *The Ecology of Human Development.* Cambridge, MA: Harvard University Press

Bronfenbrenner, U. (1986) Ecology of the Family as a Context for Human Development: Research Perspectives. *Developmental Psychology,* 22 (6): 723–742

Brown, K. and Ryan, R. (2003) The Benefits of Being Present: Mindfulness and its Role in Psychological Well-Being. *Journal of Personality and Social Psychology,* 84 (4): 822–848

Brown, K., Ryan, R. and Creswell, D. (2007) Mindfulness: Theoretical Foundations and Evidence for its Salutary Effects. *Psychological Inquiry,* 18 (4): 211–237

BTCV (2009) School Green Gym. Evaluation Findings: Health and Social Outcomes, available online at: http://www2.btcv.org.uk/display/greengym_research [accessed on 2 December 2010]

Budiansky, S. (1998) *The Nature of Horses: Their Evolution, Intelligence and Behaviour.* London: Phoenix Press

Burdette, H. and Whitaker, R. (2005) Resurrecting Free Play in Young Children: Looking Beyond Fitness and Fatness to Attention, Affiliation, and Affect. *Archives of Pediatrics and Adolescent Medicine,* 159 (1): 46–50

Burgon, H. (2003) Case Studies of Adults Receiving Horse Riding Therapy. *Anthrozoos,* 16 (3): 263–276

Butler, A., Ford, D. and Tregaskis, C. (2007) Who Do We Think We Are?: Self and Reflexivity in Social Work Practice. *Qualitative Social Work,* 6 (3): 281–299

Cahalan, W. (1995) Ecological Groundedness in Gestalt Therapy. In T. Roszak, M. Gomes and A. Kanner (Eds.), *Ecopsychology: Restoring the Earth, Healing the Mind.* San Francisco: Sierra Club Books

Cannon, W. B. (1929) *Bodily Changes in Pain, Hunger, Fear and Rage: An Account of Recent Research Into the Function of Emotional Excitement* (2nd edn.). New York, Appleton-Century-Crofts

Carns, A., Carns, M. and Holland, J. (2001) Learning the Ropes: Challenges for Change. *Therapeutic Recreation Journal,* 29: 66–71

Cawley, R., Cawley, M. and Retter, K. (1994) Therapeutic Horseback Riding and Self-Concept in Adolescents with Special Educational Needs. *Anthrozoos,* 7 (2): 129–134

Chamberlin, J. (2007) *Horse: How The Horse Has Changed Civilisations.* Oxford: Signal

Chambers, R., Gullone, E. and Allen, N. (2009) Mindful Emotion Regulation: An Integrative Review. *Clinical Psychology Review,* 29: 560–572

Chandler, C. (2005) *Animal Assisted Therapy in Counselling.* New York and Hove: Routledge

Chardonnens, E. (2009) The Use of Animals as Co-Therapists on a Farm: The Child-Horse Bond in Person-Centred Equine-Assisted Psychotherapy. *Person Centred and Experimental Psychotherapies,* 8 (4): 319–332

Chodorow, N. (1978) *The Reproduction of Mothering*. Berkeley, CA: University of California Press

Christensen, P. and James, A. (Eds.) (2000) *Research with Children: Perspectives and Practices*. London: Falmer Press

Christensen, P. and Prout, A. (2002) Working with Ethical Symmetry in Social Research with Children. *Childhood*, 9 (4): 477–497

Christensen, P. and Prout, A. (2005) Anthropological and Sociological Perspectives on the Study of Children. In S. Greene and D. Hogan (Eds.), *Researching Children's Experience: Approaches and Methods*. London: Sage

Cicchetti, D. and Rogosch, F. (1997) The Role of Self-Organization in the Promotion of Resilience in Maltreated Children. *Development and Psychopathology*, 9: 797–815

Clarke, S. (2002) Learning from Experience: Psycho-Social Research Methods in the Social Sciences. *Qualitative Research*, 2 (2): 173–194

Clarke, S., Hahn, H. and Hoggett, P. (Eds.) (2008) *Object Relations and Social Relations: The Implications of the Relational Turn In Psychoanalysis*. London: Karnac

Clarke, S. and Hoggett, P. (Eds.) (2009) *Researching Beneath the Surface: Psycho-Social Research Methods in Practice*. London: Karnac

Clarkson, P. (1999) *Gestalt Counselling in Action* (2nd edn.). London: Sage

Cleary, R. (1999) III. Bowlby's Theory of Attachment and Loss: A Feminist Reconsideration. *Feminism & Psychology*, 9 (1): 32–42

Coffey, A. and Atkinson, P. (1996) *Making Sense of Qualitative Data: Complementary Research Strategies*. Thousand Oaks, CA: Sage

Coleman, M. (2006) *Awake In The Wild: Mindfulness in Nature as a Path of Self-Discovery*. Novato, CA: New World Library

Compas, B. and Hammen, C. (1996) Child and Adolescent Depression: Covariation and Comorbidity in Development. In R. Haggerty, L. Sherrod, N. Garmezy and M. Rutter (Eds.), *Stress, Risk and Resilience in Children and Adolescents: Processes, Mechanisms, and Interventions*, Cambridge: Cambridge University Press

Compas, B., Hinden, B. and Gerhardt, C. (1995) Adolescent Development: Pathways and Processes of Risk and Resilience. *Annual Review of Psychology*, 46: 265–293

Corrigan, P. (1979) *Schooling the Smash Street Kids*. London: Macmillan

Corson, S. and Corson, E. (Eds.) (1980) *Ethology and Nonverbal Communication in Mental Health: An Interdisciplinary Biopsychosocial Exploration*. Oxford: Pergamon Press

Crawford, E., Worsham, N. and Swinehart, E. (2006) Benefits Derived From Companion Animals and the Use of the Term 'attachment'. *Anthrozoos*, 19 (2): 98–112

Crittenden, P. (1988) Distorted Patterns of Relationship in Maltreating Families: The Role of Internal Representational Models. *Journal of Reproductive and Infant Psychology*, 6 (3): 183–199

Cushing, J. and Williams, J. (1995) The Wild Mustang Programme: A Case Study in Facilitated Inmate Therapy. *Journal of Offender Rehabilitation*, 22 (3–4): 95–112

Daly, B. and Morton, L. (2006) An Investigation of Human-Animal Interactions and Empathy as Related to Pet Preference, Ownership, Attachment, and Attitudes in Children. *Anthrozoos*, 19 (2): 113–127

228   Bibliography

Daniel-McKeigue, C. (2006) Playing in the Field of Research: Creating a Bespoke Methodology to Investigate Play Therapy Practice. *British Journal of Play Therapy*, 2: 24–36

Davies, J.A. (1988) *The Reins of Life: An Instructional and Informative Manual on Riding for the Disabled* (2nd edn.). London: J.A. Allen Ltd

Davies, J. (1998) Understanding the Meanings of Children: A Reflexive Process. *Children and Society*, 12: 325–335

Davies, S. and Jones, N. A. (Eds.) (1997) *The Horse in Celtic Culture: Medieval Welsh Perspectives*. Cardiff: University of Wales Press

Davies, P., Winter, M. and Cicchetti, D. (2006) The Implications of Emotional Security Theory for Understanding and Treating Childhood Psychopathology. *Development and Psychopathology*, 18: 707–735

Day, P (1981) *Social Work and Social Control*. London: Tavistock Publications

'Cruz, H. and Jones, M. (2004) *Social Work Research: Ethical and Political Contexts*. London: Sage

DCSF (2009) Children Looked After in England (including Adoption and Care Leavers) year Ending 31 March 2009, available online at: http://www.education.gov.uk/rsgateway/DB/SFR/s000878/index.shtml [accessed on 12 January 2011]

DEFRA (2006) The Animal Welfare Act 2006, available online at: www.defra.gov.uk/animalh/welfare/act/index.htm [accessed on 25 October 2008]

Dell, C., Chalmers, D., Dell, D., Sauve, E. and MacKinnon, T. (2008) Horse as Healer: An Examination of Equine Assisted Learning in the Healing of First Nations Youth from Solvent Abuse. *Pimatisiwin: A Journal of Aboriginal and Indigenous Community Health*, 6 (1): 81–106

De Paul, J. and Guibert, M. (2008) Empathy and Child Neglect: A Theoretical Model. *Child Abuse & Neglect*, 32: 1063–1071

Denscombe, M. (2003) *The Good Research Guide for Small Scale-Scale Social Research Projects* (2nd edn.). Maidenhead: Open University Press

Denzin, N. (2001) *Interpretive Interactionism* (2nd edn.). Thousand Oaks, CA: Sage

Denzin, N. (2002) Social Work in the Seventh Moment. *Qualitative Social Work*, 1 (1): 25–38

Denzin, N. and Lincoln, Y. (2002) *The Qualitative Inquiry Reader*. Thousand Oaks, CA: Sage

Devall, B. and Sessions, G. (1985) *Deep Ecology; Living as if Nature Mattered*. Salt Lake City: Peregrine Smith Books

Dominelli, L. (1997) *Sociology for Social Work*. London: Macmillan

Donaghy, G. (2006) Equine Assisted Therapy. *Journal of Mental Health Nursing*, 26 (4): 5

Dorrance, B. and Desmond, B. (2001) *True Horsemanship Through Feel*. Guilford, CT: Lyons Press

Dossey, L. (1997) The Healing Power of Pets: A Look at Animal-Assisted Therapy. *Alternative Therapies*, 3 (4): 8–16

Dunayer, J. (2004) *Speciesism*. Derwood, MD: Ryce Publishing

Durr, P. (2008) *A Children's Environment and Health Strategy for the UK: Consultation Response*. London: The Children's Society, available online at: http://www.childrenssociety.org.uk/resources/documents/Policy/7670_full.pdf [accessed on 17 January 2011]

Bibliography 229

Ebberling, C., Pawlak, D. and Ludwig, D. (2002) Childhood Obesity; Public Health Crisis, Common Sense Cure. *Lancet*, 360: 473–482

Edney, A.T. (1995) Companion Animals and Human Health: An Overview. *Journal of the Royal Society of Medicine*, 88: 704–708

Egeland, B., Yates, T., Appleyard, K. and van Dulmen, M. (2002) The Long-Term Consequences of Maltreatment in the Early years: A Developmental Pathway Model to Antisocial Behaviour. *Children's Services: Social Policy, Research and Practice*, 5: 249–260

Ellis, A. and Greiger, R. (Eds.) (1977) *Handbook of Rational-Emotive Therapy*. New York: Springer

Ellis, J., Braff, E. and Hutchinson, S. (2001) Youth Recreation and Resiliency: Putting Theory into Practice in Fairfax County. *Therapeutic Recreation Journal*, 35 (4): 307–317

Engel, B. (1984) The Horse as a Modality for Occupational Therapy. In F. Cromwell (Ed.), *The Changing Roles of Occupational Therapists in the 1980's*. New York: Haworth Press

Engel, B. (Ed.) (1997) *Rehabilitation with the Aid of a Horse: A Collection of Studies*. Durango, CO: Barbara Engel Therapy Services

England Marketing (2009) Childhood and Nature: A Survey on Changing Relationships with Nature Access Across Generations, available online at: http://www.naturalengland.org.uk/Images/Childhood%20and%20Nature%20Survey_tcm6–10515.pdf [accessed on 24 February 2010]

Esbjourn, R. (2006) *Why Horses Heal: A Qualitative Inquiry into Equine Facilitated Psychotherapy*. Doctoral Thesis. Institute of Transpersonal Psychology, Palo Alto, CA

Erikson, E. (1995) *Childhood and Society*. London: Vintage

EST (2010) http://www.elisabethsvendsentrust.org.uk [accessed on 16 June 2010]

Evans, R. and Franklin, A. (2010) Equine Beats: Unique Rhythms (And Floating Harmony) of Horses and their Riders. In T. Edensor, (Ed.), *Geographies of Rhythm, Nature, Place, Mobility and Bodies*. Aldershot: Ashgate

Ewert, A. (1987) Research in Outdoor Adventure: Overview and Analysis. *The Bradford Papers Annual*, II: 15–28

Ewert, A., McCormick, B. and Voight, A. (2001) Outdoor Experiential Therapies: Implications for TR Practice. *Therapeutic Recreation Journal*, 35 (2): 107–22

Ewing, C., MacDonald, P., Taylor, M. and Bowers, J. (2007) Equine-Facilitated Learning for Youths with Severe Emotional Disorders: A Quantitative and Qualitative Study. *Child Youth Care Forum*, 36: 59–72

Farrell, A. (Ed.) (2005) *Ethical Research with Children*. Maidenhead: Open University Press

FCRT (2010) Fortune Centre of Riding Therapy, available online at: http://www.fortunecentre.org [accessed on 16 June 2010]

Fine, A. (2000) Animals and Therapists: Incorporating Animals in Outpatient Psychotherapy. In A. Fine (Ed.), *Handbook on Animal Assisted Therapy: Theoretical Foundations and Guidelines for Practice*. London: Academic Press

Finlay, L. and Gough, B. (2003) *Reflexivity: A Practical Guide for Researchers in Health and Social Sciences*. Oxford: Blackwell

Fook, J. (2002) *Social Work: Critical Theory and Practice*. London: Sage

230 Bibliography

Forrester, D., Kershaw, S., Moss, H. and Hughes, L. (2008) Communication Skills in Child Protection: How do Social Workers Talk to Parents? *Child and Family Social Work*, 13 (1): 41–51

Foster, A. (2009) We didn't Know Our Babies had Been Damaged by Alcohol. *The Observer*, 13 September, p. 22

Frame, D. (2006) Practices of Therapists Using Equine Facilitated/Assisted Psychotherapy in the Treatment of Adolescents Diagnosed with Depression: A Qualitative Study. Doctoral Thesis, New York University

Frank, J. D. and Frank, J. B. (1991) *Persuasion and Healing: A Comparative Study of Psychotherapy* (3rd edn.). Baltimore, MD: The John Hopkins University Press

Franklin, A. (1999) *Animals and Modern Cultures: A Sociology of Human-Animal Relations in Modernity*. London: Sage

Freud, S. (1920) *A General Introduction to Psychoanalysis*. New York: Boni & Liveright

Frewin, K. and Gardiner, B. (2005) New Age or Old Sage: A Review of Equine Assisted Psychotherapy. *Australian Journal of Counselling Psychology*, 6: 13–17

Friedmann, E., Katcher, A., Lynch, J. and Thomas, S. (1980) Animal Companions and One year Survival of Patients After Discharge from a Coronary Care Unit. *Public Health Report*, 95: 307–312

Friedmann, E., Katcher, A., Thomas, S., Lynch, J. and Messent, P. (1983) Social Interaction and Blood Pressure: Influence of Animal Companions. *The Journal of Nervous and Mental Disease*, 171 (8): 461–465

Friesen, L. (2010) Exploring Animal-Assisted Programs with Children in School and Therapeutic Contexts. *Early Childhood Journal*, 37: 261–267

Fuller, R. and Petch, A. (1995) *Practitioner Research: The Reflexive Social Worker*. Buckingham: Open University Press

Game, A. (2001) Riding: Embodying the Centaur. *Body & Society*, 7 (1): 1–12

Gammage, D. (2008) Case Study 2: Equine-Assisted Therapy. *Counselling Children and young People. BACP Quarterly Journal*, March, p. 5

Garcia, D. (2010) Of Equines and Humans: Toward a New Ecology. *Ecopsychology*, 2 (2): 85–89

Garmezy, N. (1996) Reflections and Commentary on Risk, Resilience, and Development. In R. Haggerty, L. Sherrod, N. Garmezy and M. Rutter (Eds.), *Stress, Risk and Resilience in Children and Adolescents: Processes, Mechanisms, and Interventions*. Cambridge: Cambridge University Press

Garmezy, N. and Masten, A. (1994) Chronic Adversities. In M. Rutter, L. Hersov and E. Taylor (Eds.), *Child and Adolescent Psychiatry* (3rd edn.). Oxford: Blackwell Scientific Publications

Garner, L. (2009) When a Mother and Child Must Part. *The Daily Telegraph*, 8th September, p. 22

Geertz, C. (1973) Thick Description: Toward an Interpretive Theory of Culture. In *The Interpretation of Cultures: Selected Essays*. New York: Basic Books

Gehrke, E. (2006) A Study of Heart Rate Variability (HRV) Between Horses and Humans. Paper presented to *NARHA Annual Conference*, Indianapolis, IN, 11 November

Gehrke, E., Baldwin, A. and Schiltz, P. (2011) Heart Rate Variability in Horses Engaged in Equine-Assisted Activities. *Journal of Equine Veterinary Science*, 31: 78–84

Bibliography 231

Gerhardt, S. (2004) *Why Love Matters: How Affection Shapes a Baby's Brain.* London: Routledge

Germer, C. (2005) Mindfulness: What Is It? What Does It Matter? In C. Germer, R. Siegel and P. Fulton (Eds.), *Mindfulness and Psychotherapy.* New York. The Guildford Press

Germer, C., Siegel, R. and Fulton, P. (Eds.) (2005) *Mindfulness and Psychotherapy.* New York. The Guildford Press

Gilligan, R. (1999) Enhancing the Resilience of Children and young People in Public care by Mentoring Their Talents and Interests. *Child and Family Social Work*, 4: 187–196

Gilligan, R. (2004) Promoting Resilience in Child and Family Social Work: Issues for Social Work Practice, Education and Policy. *Social Work Education*, 23 (1): 93–104

Glantz, M. and Johnson, J. (Eds.) (1999) *Resilience and Development: Positive life Adaptations.* New York: Kluwer Academic/Plenum

Goleman, D. (1996) *Emotional Intelligence: Why It Can Matter More Than IQ.* London: Bloomsbury

Goodman, T. (2005) Working with Children: Beginner's Mind. In C. Germer, R. Siegel and P. Fulton. (Eds.), *Mindfulness and Psychotherapy.* New York. The Guildford Press

Gordon, D. (1978) *Therapeutic Metaphors.* Cupertino, CA: META Publications

Graham, J (2007) An Evaluation of Equine-Assisted Wellness in those Suffering from Catastrophic Loss and Emotional Fluctuations. Doctoral Thesis. The University of Utah

Graue, E. and Walsh, D. (1998) *Studying Children in Context: Theories, Methods, and Ethics.* London: Sage

Gray, J.A. (1988) *The Psychology of Fear and Stress* (2nd edn.). Cambridge: Cambridge University Press

Green, J. and Goldwyn, R. (2002) Attachment Disorganisation and Psychopathology: New Findings in Attachment Research and Their Potential Implications for Developmental Psychopathology in Childhood. *Journal of Child Psychology and Psychiatry*, 43 (7): 835–846

Greene, S. and Hogan, D. (Eds.) (2005) *Researching Children's Experience: Approaches and Methods.* London: Sage

Grover, S. (2004) Why Won't they Listen to Us?: On Giving Power and Voice to Children Participating in Social Research. *Childhood*, 11 (1): 81–93

Gubrium, J. and Holstein, J. (1997) *The New Language of Qualitative Method.* New York: Oxford University Press

Gunner, M., Fisher, P. and The Early Experience, Stress and Prevention Network (2006) Bringing Basic Research on Early Experience and Stress Neurobiology to Bear on Preventative Interventions for Neglected and Maltreated Children. *Development and Psychopathology*, 18: 651–677

Gunner, M. and Quevedo, K. (2007) The Neurobiology of Stress and Development. *Annual Review of Psychology*, 58: 145–173

Haggerty, R., Sherrod, L., Garmezy, N. and Rutter, M. (Eds.) (1996) *Stress, Risk and Resilience in Children and Adolescents: Processes, Mechanisms, and Interventions.* Cambridge: Cambridge University Press

Hallberg, L. (2008) *Walking the Way of the Horse: Harnessing the Power of the Horse-Human Relationship.* Bloomington, IN: iUniverse

232 *Bibliography*

Halliwell, E. (2005) *Up and Running? Exercise Therapy and the Treatment of Mild or Moderate Depression in Primary Care*. London: The Mental Health Foundation

Halliwell, E. (2010) *Mindfulness Report 2010*. London: Mental Health Foundation

Hampel, P. and Petermann, F. (2006) Perceived Stress, Coping and Adjustment in Adolescents. *Journal of Adolescent Health*, 38 (4): 409–415

Hannah, B., Frantz, D. and Wintrode, A. (Eds.) (1992) *The Cat, Dog and Horse Lectures and 'The Beyond'*. Wilmette, IL: Chiron Publication

Haraway, D. (1991) A Cyborg Manifesto: Science, Technology, and Socialist-Feminism in the Late Twentieth Century. In *Simians, Cyborgs and Women: The Reinvention of Nature*. New York: Routledge

Haraway, D. (1992) *Primate Visions: Gender, Race, and Nature in the World of Modern Science*. London: Verso

Haraway, D. (2004) *The Haraway Reader*. New York: Routledge

Haraway, D. (2008) *When Species Meet*. Minnesota, MN: University of Minnesota Press

Hart, L. (2000) Psychological Benefits of Animal Companionship. In A. Fine (Ed.), *Handbook on Animal-assisted Therapy: Theoretical Foundations and Guidelines for Practice*. San Diego: Academic Press

Hayden, A. (2005) An Exploration of the Experiences of Adolescents Who Participated in Equine Faciliated Psychotherapy: A Resiliency Perspective. Doctoral Thesis. Alliant International University

Heady, B. (1999) Health Benefits and Health Cost Savings Due to Pets: Preliminary Estimates from an Australian National Survey. *Social Indicators Research*, 47 (2): 233–243

Heerwagen, J. and Orians, G. (2002) The Ecological World of Children. In P. Kahn, and S. Kellert (Eds.), *Children and Nature: Psychological, Sociocultural and Evolutionary Investigations*. Cambridge, MA: The MIT Press

Held, C. (2006) Horse Girl: An Archetypal Study of Women, Horses, And Trauma Healing. Doctoral Thesis, Pacifica Graduate Institute

Henn, M., Weinstein, M. and Foard, N. (2006) *A Short Introduction to Social Research*. London: Sage

Henrickson, J. (1971) Horseback Riding for the Handicapped. *Archives of Physical Medicine and Rehabilitation*, 52 (6): 282–283

Heppner, W. And Kernis, M. (2007) 'Quiet Ego' Functioning: The Complementary Roles of Mindfulness, Authenticity, and Secure High Self-Esteem. *Psychological Inquiry*, 18 (4): 248–251

Heppner, W., Kernis, M., Lakey, C., Keith Campbell, W., Goldman, B., Davies, P. and Cascio, E. (2008) Mindfulness as a Means of Reducing Aggressive Behavior: Dispositional and Situational Evidence. *Aggressive Behavior*, 34 (5): 486–496

Hertz, R. (Ed.) (1977) *Reflexivity and Voice*. Thousand Oaks, CA: Sage

Hingley-Jones, H. (2009) Developing Practice-Near Social Work Research to Explore the Emotional Worlds of Severely Learning Disabled Adolescents in 'Transition' and their Families. *Journal of Social Work Practice*, 23 (4): 413–426

HM Prison Service (2004) Horses for Courses. *Prison Service News*, available online at: http://www.hmprisonservice.gov.uk/prisoninformation/prison servicemagazine/index.asp?id=3876,18,3,18,0,0 [accessed on 16 June 2010]

Hoggett, P. (2008) Relational Thinking and Welfare Practice. In S. Clarke, H. Hahn and P. Hoggett (Eds.), *Object Relations and Social Relations: The Implications of the Relational Turn in Psychoanalysis*. London: Karnac

Bibliography 233

Holland, S. (2001) Representing Children in Child Protection. *Childhood*, 8 (3): 322–339

Holland, S., Renold, E., Ross, N. and Hillman, R. (2010) Power, Agency and Participatory Agendas: A Critical Exploration of Young People's Engagement in Participative Qualitative Research. *Childhood*, 17 (3): 360–375

Hollway, W. (2008) The Importance of Relational Thinking in the Practice of Psycho-Social Research: Ontology, Epistemology, Methodology, and Ethics. In S. Clarke, H. Hahn, and P. Hoggett (Eds.), *Object Relations and Social Relations: The Implications of The Relational Turn In Psychoanalysis*. London: Karnac

Hollway, W. (2009) Applying The 'Experience-Near' Principle to Research: Psychoanalytically Informed Methods. *Journal of Social Work Practice*, 23 (4): 461–474

Hollway, W. and Jefferson, T. (2000) *Doing Qualitative Research Differently: Free Association, Narrative and the Interview Method*. London: Sage

Holmes, J. (1993) *John Bowlby and Attachment Theory*. London: Routledge

Hooker, S., Freeman, L. and Stewart, P. (2002) Pet Therapy Research: A Historical Review. *Holistic Nursing Practice*, 17 (1): 17–23

Hoskins, M. and Stoltz, J.A. (2005) Fear of Offending: Disclosing Researcher Discomfort When Engaging in Analysis. *Qualitative Research*, 5 (1): 95–111

Houpt, K., Law, K. and Martinisi, V. (1978) Dominance Hierarchies in Domestic Horses. *Applied Animal Ethology*, 4 (3): 273–83

Howard, S., Dryden, J. and Johnson, B. (1999) Childhood Resilience: Review and Critique of Literature. *Oxford Review of Education*, 25 (3): 308–323

Howard, S. and Johnson, B. (2000) What Makes the Difference? Children and Teachers Talk About Resilient Outcomes for Children 'at risk'. *Educational Studies*, 26 (3): 321–337

Howe, D. (2005) *Child Abuse and Neglect: Attachment, Development and Intervention*. New York: Palgrave MacMillan

Howe, D. (2009) *A Brief Introduction to Social Work Theory*. Basingstoke, Hants: Palgrave MacMillan

Howe, D. and Fearnley, S. (1999) Disorders of Attachment and Attachment Therapy. *Adoption & Fostering*, 23 (2): 19–31

Howey, M. (2002) *The Horse in Magic and Myth*. New York: Dover Publications, Inc

Hughes, D. (2004) *Faciliating Developmental Attachment: The Road to Emotional Recovery and Behavioural Change in Foster and Adopted Children*. Lanham, MD: Rowman & Littlefield Inc

Humphries, B. (1997) From Critical Thought to Emancipatory Action: Contradictory Research Goals? *Sociological Research Online*, 2 (1), available online at: http://socresonline.org.uk/socresonline/2/1/3.html [accessed on 27 October 2006]

Hyland, A. (1996) *The Medieval Warhorse from Byzantium to the Crusades*. Stroud, Glos: Sutton Publishing

Illich, I. (1975) *Medical Nemesis: The Expropriation of Health*. London: Calder and Boyars

Irvine, L. (2004) *If You Tame Me: Understanding our Connection with Animals*. Philadelphia: Temple University Press

234 *Bibliography*

Irwin, C. (2005) *Dancing with your Dark Horse: How Horse Strength Helps Us Find Balance, Strength And Wisdom.* New York: Marlowe and Company

Isaacson, R. (2009) *Horse Boy: A Father's Miraculous Journey to Heal His Son.* London: Viking/Penguin Books

Jack, G. (2010) Place Matters: The Significance of Place Attachments for Children's Well-Being. *British Journal of Social Work*, 40 (3): 755–771

Jackson, S. (2006) *The Horse in Myth and Legend.* Stroud, Glos: Tempus Publishing Ltd

Jackson, S. and McParlin, P. (2006) The Education of Children in Care. *The Psychologist*, 19 (2): 90–93

Jackson, S. and Simon, A. (2005) The Costs and Benefits of Educating Children in Care. In E. Chase, A. Simon and S. Jackson (Eds.), *In Care and After: A Positive Perspective.* London: Routledge

Jackson, S., Born, M. and Jacob, M-N. (1997) Reflections on Risk and Resilience in Adolescence. *Journal of Adolescence*, 20 (6): 609–616

James, A. and Prout, A. (Eds.) (1990) *Constructing and Reconstructing Childhood.* London: Falmer Press

Jones, A. (2004) Involving Children and Young People as Researchers. In S. Fraser, V. Lewis, S. Ding, M. Kellett and C. Robinson (Eds.), *Doing Research with Children and Young People.* London: Sage

Jones, B. (1983) Just Crazy About Horses: The Fact Behind the Fiction. In A. Katcher and A. Beck (Eds.), *New Perspectives on Our Lives with Companion Animals.* Philadelphia, PA: University of Pennsylvania Press

Jones, P. (1996) *Drama as Therapy: Theatre as Living.* London: Routledge

Jones, R. (2003) The Construction of Emotional and Behavioural Difficulties. *Educational Psychology in Practice*, 19 (2):147157

Jones, R. (2008) Storytelling, Socialization and Individuation. In R. Jones, A. Clarkson, S. Congram, and N. Stratton (Eds.), *Education and Imagination: Post-Jungian Perspectives.* London: Routledge

Jones, R., Clarkson, A., Congram, S. and Stratton, N. (Eds.) (2008) *Education and Imagination: Post-Jungian Perspectives.* London: Routledge

Jung, C. G. (Ed.) (1978) *Man and his Symbols.* London: Picador

Jung, C.G. (1983) *Memories, Dreams, Reflections.* London: Flamingo

Jung, C. G. (1990) *The Archetypes and the Collective Unconscious.* London: Routledge (Original work published 1959)

Kabat-Zinn, J. (1994) *Wherever You Go, There You Are: Mindfulness Meditation for Everyday Life.* London: Piatkus Books Ltd

Kabat-Zinn, J., Massion, A., Kristeller, J., Peterson, K., Fletcher, K. Pbert, L., Lenderking, W. and Santorelli, S. (1992) Effectiveness of a Meditation-based Stress Reduction Program in the Treatment of Anxiety Disorders. *American Journal of Psychiatry*, 149: 936–943

Kahn, P. and Kellert, S. (Eds.) (2002) *Children and Nature: Psychological, Sociocultural and Evolutionary Investigations.* Cambridge, MA: The MIT Press

Kaiser, L., Smith, K., Heleski, C. and Spence, L. (2006) Effects of a Therapeutic Riding Program on at-Risk and Special Educational Children. *Journal of American Medical Association*, 228 (1): 46–52

Kaiser, L., Spence, J., Lavergne, A. And Bosch, K. (2004) Can a Week of Therapeutic Riding Make a Difference? – A Pilot Study. *Anthrozoos*, 17 (1): 63–72

Bibliography 235

Kaplan, S. (1995) The Restorative Benefits of Nature: Toward an Integrative Framework. *Journal of Environmental Psychology*, 15: 169–182

Karen, R. (1998) *Becoming Attached: First Relationships and How They Shape Our Capacity to Love.* Oxford: Oxford University Press

Karol, J. (2007) Applying a Traditional Individual Psychotherapy Model to Equine-Facilitated Psychotherapy (EFP): Theory and Method. *Clinical Child Psychology and Psychiatry*, 12: 77–90

Karol, J. (2007) Applying a Traditional Individual Psychotherapy Model to Equine-facilitated Psychotherapy (EFP): Theory and Method. *Clinical Child Psychology and Psychiatry*, 12: 77–90

Katcher, A. and Wilkins, G. (2000) The Centaur's Lessons: Therapeutic Education Through Care of Animals and Nature Study. In A. Fine (Ed.), *Handbook on Animal Assisted Therapy: Theoretical Foundations and Guidelines for Practice.* San Diego, CA: Academic Press

Katz, M. (1997) *On Playing a Poor Hand Well. Insights from the Lives of those Who have Overcome Childhood Risks and Adversities.* New York: W.W. Norton & Company, Inc

Kellert, S. (2002) Experiencing Nature: Affective, Cognitive, and Evaluative Development in Children. In P. Kahn and S. Kellert (Eds.), *Children and Nature: Psychological, Sociocultural and Evolutionary Investigations*, Cambridge, MA: The MIT Press

Kellert, S. and Wilson, E. (Eds.) (1993) *The Biophilia Hypothesis.* Washington: Island Press

Kemp, M. (2007) *The Human Animal History Today*, November: 34–41

Kemp, K., Signal, T., Botros, H., Taylor, N. and Prentice, K. (2013) Equine Facilitated Therapy with Children and Adolescents Who Have Been Sexually Abused: a Program Evaluation Study. *Journal of Child and Family Studies*, available online at: http://dx.doi.org/10.1007/s10826-013-9718-1 [accessed on 24 January 2013]

Kiley-Worthington, M. (1987) *The Behaviour of Horses: In Relation to Management and Training.* London: J.A. Allen

Kimball, C. (2002) *Mindful Horsemanship: Increasing Your Awareness One Day at a Time.* Middleton, New Hampshire: Carriage House Publishing

Klein, M. (1975) *The Psycho-Analysis of Children.* London: Virago

Klontz, B., Bivens, A., Leinart, D. and Klontz, T. (2007) The Effectiveness of Equine-Assisted Experiential Therapy: Results of an Open Clinical Trial. *Society and Animals*, 15: 257–267

Knox, J. (2003) *Archetype, Attachment, Analysis: Jungian Psychology and the Emergent Mind.* East Sussex: Brunner-Routledge

Kogan, L., Granger, B., Fitchett, J., Helmer, K. and Young, K. (1999) The Human-animal Team Approach for Children with Emotional Disorders: Two Case Studies. *Child & Youth Care Forum*, 28 (2): 105–121

Kohanov, L. (2001) *The Tao of Equus: A Woman's Journey of Healing and Transformation Through the Way of the Horse.* Novato, CA: New World Library

Kohanov, L. (2005) *Riding Between the Worlds: Expanding Our Potential Through the Way of the Horse.* Novato, CA: New World Library

Korpela, K., Hartig, T., Kaiser, F. and Fuhrer, U. (2001) Restorative Experience and Self-Regulation in Favorite Places. *Environment and Behavior*, 33: 572–589

236   *Bibliography*

Kraemer, J., Tebes, J., Kaufman, J., Adnopoz, J. and Racusin, G. (2001) Resilience and Family Psychosocial Processes among Children of Parents with Serious Mental Disorders. *Journal of Child and Family Studies*, 10 (1): 115–136

Krane, J., Davies, L., Carlton, R. and Mulcahy, M. (2009) The Clock Starts Now: Feminism, Mothering and Attachment Theory in Child Protection Practice. In B. Featherstone, C.-A. Hooper, J. Scourfield and J. Taylor (Eds.), *Gender and Child Welfare in Society*. Chichester: Wiley Blackwell

Kunzle, U., Steinlin, E. and Yasikoff, N. (1994) 'Hippotherapy-K': The Healing Rhythmical Movement of the Horse for Patients with Multiple Sclerosis. In P. Eaton (Ed.), *Eighth International Therapeutic Riding Congress: The Complete Papers*. New Zealand: National Training Resource Centre

Kurtz, R. (1990) *Body-Centred Psychotherapy: The Hakomi Method*. Mendocino, CA: LifeRhythm

Kuyken, W., Byford, S., Taylor, R. S., Watkins, E., Holden, E., White, K., Barrett, B., Byng, R., Evans, A., Mullan, E. and Teasdale, J. D. (2008) Mindfulness-based Cognitive Therapy to Prevent Relapse in Recurrent Depression. *Journal of Consulting Clinical Psychology*, 76 (6): 966–978

Laird, J. (1994) 'Thick Description' Revisited: Family Therapist as Anthropologist-Constructivist. In E. Sherman and W. Reid (Eds.), *Qualitative Research in Social Work*. New York: Columbia University Press

Laing, R. D. (1968) *The Politics of Experience and the Bird of Paradise*. Harmondsworth: Penguin

Lambert, M. and Bergin, A. (1994) The Effectiveness of Psychotherapy. In A. Bergin and S. Garfield (Eds.), *Handbook of Psychotherapy and Behavior Change* (4th edn.). New York: Wiley

Latimer, J. and Birke, L. (2009) Natural Relations: Horses, Knowledge and Technology. *The Sociological Review*, 57 (1): 1–27

Lawrence, E. (1984) *Rodeo: An Anthropologist Looks at the Wild and the Tame*. Chicago: University of Chicago Press

Lazarus, R. (1966) *Psychological Stress and the Coping Process*. New York: Mc-Graw-Hill

Lazarus, R. (1993) From Psychological Stress to Emotions: A History of Changing Outlooks. *Annual Review of Psychology*, 44: 1–21

Lazarus, R. and Folkman, S. (1984) *Stress, Appraisal and Coping*. New York: Springer

LEAP (2010) http://www.leap-etc.co.uk [accessed on 16 June 2010]

Lentini, J. and Knox, M. (2009) A Qualitative and Quantitative Review of Equine Facilitated Psychotherapy (EFP) with Children and Adolescents. *The Open Complementary Medicine Journal*, 1: 51–57

Leupnitz, D. (1988) *The Family Interpreted*. New York: Basic Books

Levine, P. (1997) *Waking the Tiger: The Innate Capacity to Transform Overwhelming Experiences*. Berkeley, CA: North Atlantic Books

Levine, P. (2005) *Healing Trauma: A Pioneering Program for Restoring the Wisdom of your Body*. Boulder, CO: Sounds True

Levinson, B. (1969) *Pet-oriented Child Psychotherapy*. Springfield, IL: Charles C Thomas

Levinson, B. (1972) *Pets and Human Development*. Springfield, IL: Charles C Thomas

Levinson, B. (1980) The Child and His Pet: A World of Nonverbal Communication. In S. Corson, and E. Corson (Eds.), *Ethology and Nonverbal*

Communication in Mental Health: An Interdisciplinary Biopsychosocial Exploration. Oxford: Pergamon Press

Levinson, D., Darrow, C., Klein, E., Levinson, M. and McKee, B. (1978) *The Seasons of a Man's Life.* New York: Knopf

Levi-Strauss, C. (1968) *Totemism.* Harmondsworth: Penguin

Lexman, J. and Reeves, R. (2009) *Building Character.* London: Demos Report, available online at: www.demos.co.uk [accessed on 27 April 2010]

Linden, S. and Grut, J. (2002) *The Healing Fields: Working with Psychotherapy and Nature to Rebuild Shattered Lives.* London: Francis Lincoln

Llewellyn, A. (Ed.) (2007) *The Healing Touch of Horses: True Stories of Courage, Hope and Transformative Power of the Human/Equine Bond.* Avon, MA: Adams Media

Loch, S. (1999) *The Royal Horse of Europe: The Story of the Andalusian and Lusitano.* London: JA Allen

Louv, R. (2008) *Last Child in the Woods: Saving Our Children From Nature-Deficit Disorder* (2nd edn.). Alonquin: Chapel Hill

Lovelock, J. (1979) *Gaia: A New Look at Life on Earth.* New York: Oxford University Press

Luther, S., Cicchetti, D. and Becker, B. (2000) The Construct of Resilience: A Critical Evaluation and Guidelines for Future Work. *Child Development,* 71 (3): 543–562

Lyons-Ruth, K. (2003) Dissociation and the Parent-infant Dialogue: A Longitudinal Perspective from Attachment Research. *Journal of the American Psychoanalytic Association,* 51 (3): 883–911

Lynch, R. and Garrett, P. (2010) 'More than Words': Touch Practices in Child and Family Social Work. *Child and Family Social Work,* 15: 389–398

Mackewn, J. (1997) *Developing Gestalt Counselling: A Field Theoretical and Relational Model of Contemporary Gestalt Counselling and Psychotherapy.* London: Sage

MacKinnon, J., Noh, S., Laliberte, J. and Allan, D. (1995) Therapeutic Horseback Riding: A Review of the Literature. *Physical and Occupational Therapy in Paediatrics,* 15 (1): 1–15

MacSwiney of Mashanaglass, The Marquis, O. (1987) *Training from the Ground: A Special Approach.* London: J.A. Allen & Co

Mallon, G. (1994) Cow as Co-Therapist: Utilization of Farm Animals as Therapeutic Aids with Children in Residential Treatment. *Child and Adolescent Social Work Journal,* 11 (6): 455–474

March, J. (2009) The Future of Psychotherapy for Mentally Ill Children and Adolescents. *Journal of Child Psychology and Psychiatry,* 50 (1–2): 170–179

Marr, C., French, L., Thompson, D., Drum, L., Greening, G., Mormon, J., Henderson, I. and Huges, C. (2000) Animal-assisted Therapy in Psychiatric Rehabilitation. *Anthrozoos,* 13 (1): 43–37

Maslow, A. (1968) *Towards a Psychology of Being* (2nd edn.). New York: D. Van Nostrand Company

Mason, O. and Hargreaves, I. (2001) A Qualitative Study of Mindfulness-Based Cognitive Therapy for Depression. *British Journal of Medical Psychology,* 74: 197–212

Masten, A., Best, K. and Garmezy, N. (1990) Resilence and Development: Contributions From the Study of Children Who Overcome Adversity. *Development and Psychopathology,* 2 (4): 425–444

238   *Bibliography*

Masten, A. and Obradovic, J. (2006) Competence and Resilience in Development. *Annals of the New York Academy of Science*, 1094: 13–27

Mayberry, R (1978) The Mystique of the Horse is Strong Medicine: Riding as Therapeutic Recreation. *Rehabilitation Literature*, 38 (6): 192–196

May, T. (2001) *Social Research: Issues, Methods and Process* (3rd edn.). Buckingham: Open University Press

Mazis, G. (2008) *Humans, Animals, Machines: Blurring Boundaries*. Albany: State University of New York Press

McCormick, A. and McCormick, M. (1997) *Horse Sense and the Human Heart: What Horses Can Teach Us About Trust, Bonding, Creativity and Spirituality*. Florida: Health Communications Inc

McCormick, A., McCormick, M. and McCormick, T. (2004) *Horses and the Mystical Path: The Celtic Way of Expanding the Human Soul*. Novato, CA: New World Library

McCurdy, L., Winterbottom, K., Mehta, S. and Roberts, J. (2010) Using Nature and Outdoor Activity to Improve Children's Health. *Current Problems in Pediatric and Adolescent Health Care*, 40 (5): 102–117

McGreevy, P. (2004) *Horse Behaviour*. London: Elsevier Saunders

McLeod, J. (2001) *Qualitative Research in Counselling and Psychotherapy*. London: Sage

McLoughlin, B. (1995) *Developing Psychodynamic Counselling*. London: Sage

McLaughlin (2005) Young Service Users as Co-researchers thodological Problems and Possibilities. *Qualitative Social Work*,): 211–228

McVeigh, T. (2009) Take More Babies Away From Bad Parents. *The Observer*, Sunday 6th September, pp. 1–5

Mearns, D. (1994) *Developing Person-Centred Counselling*. London: Sage

Mearns, D. and Thorne, B. (2000) *Person Centred Therapy Today: New Frontiers in Theory and Practice*. London: Sage

Meinersmann, K., Bradberry, J. and Bright Roberts, F. (2008) Equine-Facilitated Psychotherapy with Adult Female Survivors of Abuse. *Journal of Psychosocial Nursing*, 46 (12): 36–42

Melson, G. (1990) Studying Children's Attachments to their Pets: A Conceptual and Methodological Review. *Anthrozoos*, 4 (2): 91–99

Melson, G. (2001) *Why The Wild Things Are: Animals in the Lives of Children*. Cambridge, MA: Harvard University Press

Melson, G. (2003) Child Development and the Human-Companion Animal Bond. *American Behavioral Scientist*, 47 (1): 31–39

Merz-Perez, L. and Heide, K. (2004) *Animal Cruelty: Pathway to Violence Against People*. New York: Altamira Press

MIND (2007) Ecotherapy: The Green Agenda for Mental Health. Executive Summary, available online at: http://www.mind.org.uk/help/ecominds/ecominds/mental_health_and_the_environment [accessed on 15 January 2011]

Misra, M., Pacaud, D., Petryk, A., Collett-Solberg, P. F. and Kappy, M. (2008) Vitamin D Deficiency in Children and Its Management: Review of Current Knowledge and Recommendations. *Pediatrics*, 122 (2): 398–417

Mistral, K. (2007) Heart to Heart: A Quantitative Approach to Measuring the Emotional Bond between Horses and Humans. *Horse Connection Magazine*, August, pp. 44–47

Moffitt, T. (1993) Adolescence-Limited and Life-Course-Persistent Antisocial Behaviour: A Developmental Taxonomy. *Psychological Review*, 100: 674–701

Moorhead, J. (2010) Not a Cure, a Healing. *The Guardian*, Family Section, Saturday 23 January, p. 2

Moote, G. and Wodarski, J. (1997) The Acquisition of Life Skills Through Adventure-Based Activities and Programs: A Review of the Literature. *Adolescence*, 32 (125): 143–163

Moreau, L. (2001) Outlaw Riders: Equine-facilitated Therapy with Juvenile Capital Offenders. *Reaching Today's Youth*, 5 (2): 27–30, available online at: http://www. cyc-net.org/cyc-online/cycol-0405-outlawriders.html [accessed on 24 February 2008]

Morgan, P. (2010) 'Get Up. Stand Up': Riding to Resilience on a Surfboard. *Child and Family Social Work*, 15 (1): 56–65

Morgan, W. and Morgan, S. (2005) Cultivating Attention and Empathy. In C. Germer, R. Siegel and P. Fulton (Eds.), *Mindfulness and Psychotherapy*. New York: The Guildford Press

Morrow, V. and Richards, M. (1996) The Ethics of Social Research with Children: An Overview. *Children and Society*, 10 (2): 90–105

Moss, D., Waugh, M. and Barnes, R. K. (2008) A Tool for Life? Mindfulness as Self Help or Safe Uncertainty. *International Journal of Qualitative Studies on Health and Well-being*, 3 (3): 132–142

Moustakas, C. (1990) *Heuristic Research: Design, Methodology and Applications*. London: Sage

Muñoz, S. (2009) *Children in the Outdoors: A Literature Review*. Forres, Scotland, Sustainable Development Research Centre, available online at: http:// www.countrysiderecreation.org.uk/Children%20Outdoors.pdf [accessed on 17 January 2011]

Murray, J., Browne, W., Roberts, M., Whitmarch, A. and Gruffydd-Jones, T. (2010) Number and Ownership Profiles of Cats and Dogs in the UK. *Veterinary Record*, 166 (6): 163–168

Myers, G. (2007) *The Significance of Children and Animals: Social Development and Our Connections to Other Species* (2nd Revised edn.). West Lafayette: Purdue University Press

National Children's Bureau (2003) *Guidelines for Research*, available online at: www.ncb.org.uk [accessed on 4 October 2007]

Nebbe, L. (2000) Nature Therapy. In A. Fine (Ed.), *Handbook on Animal-Assisted Therapy: Theoretical Foundations and Guidelines for Practice*. San Diego, CA: Academic Press

Neckoway, R., Brownlee, K. and Castellan, B. (2007) Is Attachment Theory Consistent with Aboriginal Realities?. *First Peoples Child & Family Review*, 3 (2): 65–74

Netting, F., Wilson, C. and New, J. (1987) The Human-Animal Bond: Implications for Practice. *Social Work*, January–February, pp. 60–64

Nettles, S. and Pleck, J. (1996) Risk, Resilience, and Development: The Multiple Ecologies of Black Adolescents in the United States. In R. Haggerty, L. Sherrod, N. Garmezy and M. Rutter (Eds.), *Stress, Risk and Resilience in Children and Adolescents: Processes, Mechanisms, and Interventions*, Cambridge: Cambridge University Press

240 *Bibliography*

O'Brien, L. and Murray, R. (2005) Forest Schools in England and Wales: Woodland Space to Learn and Grow. *Environmental Education*, Autumn, pp. 25–27

O'Hara, M. (1986) Heuristic Inquiry as Psychotherapy: The Client-Centered Approach. *Person-Centered Review*, 1 (2): 172–184

Okely, J. (1983) *The Traveller-Gypsies*. London: Cambridge University Press

O'Rourke, K. (2004) Horse-assisted Therapy: Good for Humans, But What About the Horses?. *Journal of the American Veterinary Association*, 204: 131–133

Parish-Plass, N. (2008) Animal-assisted Therapy with Children Suffering from Insecure Attachment Due to Abuse and Neglect: A Method to Lower the Risk of Intergenerational Transmission of Abuse? *Clinical Child Psychiatry*, 13 (1): 7–30

PAT (2010) http://www.petsastherapy.org [accessed on 16 June 2010]

Patton, M. (1990) *Qualitative Evaluation and Research Methods* (2nd edn.). Newbury Park, CA: Sage

Paul, E. And Serpell, J. (1993) Childhood Pet Keeping and Humane Attitudes in Young Adulthood. *Animal Welfare*, 2: 321–337

Paul, E. and Serpell, J. (2000) Empathy with Animals and with Humans: Are They Linked? *Anthrozoos*, 13 (4): 194–202

Payne, G. and Williams, M. (2005) *Generalization in Qualitative Research*, 39 (2): 295–314

Peacock, J., Hine, R. and Pretty, J. (2007) *Ecotherapy: The Green Agenda for Mental Health*. London: MIND, available online at: http://www.mind.org.uk/help/ecominds/ecominds/mental_health_and_the_environment [accessed on 15 January 2011]

Pemberton, C. (2010) Should Children's Social Work be a Touch Free Zone. *Community Care*, 26 August, available online at: http://www.communitycare.co.uk/Articles/2010/08/20/115114/Should-children39s-social-work-be-a-touch-free-zone.htm [accessed on 08 November 2010]

Perls, F., Hefferline, R. and Goodman, P. (1972) *Gestalt Therapy: Excitement and Growth in the Human Personality*. London: Souvenir Press

Philo, C. and Wilbert, C. (2000) *Animal Spaces, Beastly Places*. London: Routledge

Place, M., Reynolds, J., Cousins, A. and O'Neill, S. (2002) Developing a Resilience Package for Vulnerable Children. *Child and Adolescent Mental Health*, 7 (4): 162–167

Poresky, R. H. (1990) The young children's empathy measure: Reliability, validity and effects of companion animal bonding. *Psychological Reports*, 65 (5): 931936

Price, J. and Glad, K. (2003) Hostile Attributional Tendencies in Maltreated Children. *Journal of Abnormal Child Psychology*, 31: 329–343

Priest, P. (2007) Using a Walking Group to Feel Better. *Journal of Health Psychology*, 12 (1): 36–52

Podberscek, A., Paul, E. and Serpell, J. (Eds.) (2000) *Companion Animals and Us: Exploring the Relationships Between People and Pets*. Cambridge: Cambridge University Press

Poresky, R. H. (1990) The Young Children's Empathy Measure: Reliability, Validity and Effects of Companion Animal Bonding. *Psychological Reports*, 65 (5): 931–936

Pugsley, L. (2002) Putting your Oar in: Moulding, Muddling or Meddling?. In T. Welland and L. Pugsley (Eds.), *Ethical Dilemmas in Qualitative Research*. Aldershot: Ashgate

Rainer, P., Waltner-Toews, D., Bonnett, B., Woodward, C. and Abernathy, T. (1999) Influence of Companion Animals on the Physical and Psychological Health of Older People: An Analysis of a One-year Longitudinal Study. *Journal of the American Geriatrics Society*, 47 (3): 323–323

Rashid, M. (2004) *Horses Never Lie: The Heart of Passive Leadership*. Newton Abbot, Devon: David & Charles

Rector, B. (2005) *Adventures in Awareness: Learning with the Help of Horses*. Bloomington, IN: AuthorHouse

Rees, L. (1984) *The Horse's Mind*. London: Stanley Paul

Rees, L. (2009/2010) Equine Ethologist and Author (Personal Communication)

Regan, T. (1983) *The Case for Animal Rights*. Berkeley: University of California Press

Rew, L. (2000) Friends and Pets as Companions: Strategies for Coping with Loneliness Among Homeless Youth. *Journal of Child and Adolescent Psychiatric Nursing*, 13 (3): 125–140

Richards, S. (2006) *Chosen by a Horse (A Memoir)*. New York: Soho Press

Riessman, C. (Ed.) (1994) *Qualitative Studies in Social Work Research*. Thousand Oaks, CA: Sage

Risley-Curtiss, C. (2010) Social Work Practitioners and the Human-Companion Animal Bond: A National Study. *Social Work*, 55 (1): 38–46

Roberts, M. (2000) *Horse Sense for People*. London: HarperCollins

Roberts, F., Bradberry, J. and Williams, C. (2004) Equine-facilitated Psychotherapy Benefits Students and Children. *Holistic Nursing Practice*, 18 (1): 32–35

Robinson, C. and Kellett, M. (2004) Power. In S. Fraser, V. Lewis, S. Ding, M. Kellett and C. Robinson (Eds.), *Doing Research with Children and Young People*. London: Sage

Rogers, C. (1951) *Client-Centred Therapy*. Boston: Houghton Mifflin

Rogers, C. (1967) On Becoming a Person: A Therapist's View of Psychotherapy. London: Constable and Constable

Rogers, C. (1980) *A Way of Being*. Boston: Houghton Mifflin

Rogers, C. (1992) The Necessary and Sufficient Conditions of Therapeutic Personality Change. *Journal of Consulting and Clinical Psychology*, 60 (6): 827–832

Rolfe, J. (2007) *Ride from the Heart: The Art of Communication between Horse and Rider*. London: JA Allen

Rosenthal, S. R. (1975) Risk Exercise and the Physically Handicapped. *Rehabilitation Literature*, 36 (5): 144–148

Ross, N., Renold, E., Holland, S. and Hillman, A. (2009) Moving Stories: Using Mobile Methods to Explore the Everyday lives of Young People in Public Care. *Qualitative Research*, 9 (5): 605–623

Rossman, G. and Rallis, S. (1998) *Learning in the Field: An Introduction to Qualitative Research*. Thousand Oaks, CA: Sage

Roszak, T., Gomes, M. and Kanner, A. (Eds.) (1995) *Ecopsychology: Restoring the Earth, Healing the Mind*. San Francisco: Sierra Club Books

Rothbaum, F., Weisz, J., Pott, M., Miyake, K. and Gilda, M. (2000) Attachment and Culture: Security in the United States and Japan. *American Psychologist*, 55 (10): 1093–1104

Rothe, E., Vega, B., Torres, R., Soler, S. and Pazos, R. (2005) From Kids and Horses: Equine Facilitated Psychotherapy for Children. *International Journal of Clinical and Health Psychology*, 5 (2): 373–383

Runnquist, A. (Ed.) (1957) *Horses in Fact and Fiction*. London: Jonathan Cape

Rutter, M. (1979) Protective Factors in Children's Responses to Stress and Disadvantage. In M. Kent and J. Rolf (Eds.), *Primary Prevention of Psychopathology:*

242 *Bibliography*

Vol.3. Social Competence in Children. Hanover, NH: University Press of New England

Rutter, M. (1972) *Maternal Deprivation Reassessed.* Harmondsworth: Penguin Books

Rutter, M. (1985) Resilience in the Face of Adversity: Protective Factors and Resistance to Psychiatric Disorder. *British Journal of Psychiatry,* 147: 598–611

Rutter, M. (1999) Resilience Concepts and Findings: Implications for Family Therapy. *Journal of Family Therapy,* 21: 119–144

Rutter, M. (2006) Implications of Resilience Concepts for Scientific Understanding. *Annals of the New York Academy of Science,* 1094: 1–12

Rutter, M., Giller, H. and Hagell, A. (1998) *Antisocial Behaviour by Young People.* New York: Cambridge University Press

Rycroft, C. (1972) *A Critical Dictionary of Psychoanalysis.* Harmondsworth, Middlesex: Penguin

Sable, P (1995) Pets, Attachment, and Well-Being across the Life Cycle. *Social Work,* 40 (3): 334–341

Scanlan, L. (2001) *Wild about Horses: Our Timeless Passion for the Horse.* New York: Perennial

Scarr, S. and McCartney, K. (1983) How People Make Their Own Environments: A Theory of Genotype Environmental Effects. *Child Development,* 5: 424–435

SCAS (2010) http://www.scas.org.uk [accessed on 16 June 2010]

Schultz, P., Remick-Barlow, G. and Robbins, L. (2007) Equine-assisted Psychotherapy: A Mental Health Promotion/Intervention Modality for Children Who Have Experienced Intra-family Violence. *Health and Social Care in the Community,* 15 (3): 265–271

Schwartz, C. and Sendor, R. (1999) Helping Others Helps Oneself: Response Shift Effects in Peer Support. *Social Science and Medicine,* 48: 1563–1575

Scott, J. (1980) Nonverbal Communication in the Process of Social Attachment. In S. Corson and E. Corson (Eds.), *Ethology and Nonverbal Communication in Mental Health: An Interdisciplinary Biopsychosocial Exploration.* Oxford: Pergamon Press

Sedgewick, D. (2001) *Introduction to Jungian Psychotherapy: The Therapeutic Relationship.* London: Routledge

Segal, Z., Williams, J. and Teasdale, J. (2002) *Mindfulness Based Cognitive Therapy for Depression: A New Approach to Preventing Relapse.* London: Guildford Press

Selby, A and Smith-Osborne, A. (2013) A Systematic Review of Effectiveness of Complementary and Adjunct Therapies and Interventions Involving Equines. *Health Psychology,* 32 (4): 418–432

Sempik, J. Aldridge, J. and Becker, S. (2005) *Health, Well-Being and Social Inclusion: Therapeutic Horticulture in the UK.* Bristol: Polity Press

Serpell, J. (1991) Beneficial Effects of Pet Ownership on Some Aspects of Human Health and Behaviour. *Journal of the Royal Society of Medicine,* 84 (12): 717–720

Serpell, J. (1999) Guest Editor's Introduction: Animals in Children's Lives. *Society and Animals,* 7 (2): 87–94

Serpell, J. (2000a) Animal Companions and Human Well-Being: An Historical Exploration of the Value of Human-Animal Relationships. In A. Fine (Ed.), *Handbook on Animal-Assisted Therapy: Theoretical Foundations and Guidelines for Practice.* San Diego, CA: Academic Press

Bibliography 243

Serpell, J. (2000b) Creatures of the Unconscious: Companion Animals as Mediators. In A. Podberscek, E. Paul and J. Serpell (Eds.), *Companion Animals and Us: Exploring the Relationships Between People and Pets*, Cambridge: Cambridge University Press

Seyle, H. (1950) *Stress: The Physiology and Pathology of Exposure to Stress*. Montreal, QC: Acta Medical Publishers

Shaver, P., Lavy, S., Saron, C. and Mikulincer, M. (2007) Social Foundations of the Capacity for Mindfulness: An Attachment Perspective. *Psychological Inquiry*, 18 (4): 264–271

Shepard, P. (1982) *Nature and Madness*. San Francisco: Sierra Club Books

Shepard, P. (1996) *Traces of an Omnivore*. Washington DC: Island Press/Shearwater Books

Sibley, D. (1995) *Geographies of Exclusion: Society and Difference in the West*. London: Routledge

Siegel, J. (1993) Companion Animals: In Sickness and in Health. *Journal of Social Issues*, 49: 157–167

Sills, F. (2009) *Being and Becoming: Psychodynamics, Buddhism and the Origins of Selfhood*. Berkeley, CA: North Atlantic Books

Silverman, D. (2005) *Doing Qualitative Research* (2nd edn.). London: Sage

Singer, P. (1990) *Animal Liberation* (2nd edn.). New York: New York Review of Books

Skinner, E. and Zimmer-Gembeck, M. (2007) The Development of Coping. *Annual Review of Psychology*, 58: 119–144

Sloane, R., Staples, F., Cristol, A., Yorkston, N. and Whipple, K. (1975) *Psychotherapy versus Behavior Therapy*. Cambridge: Harvard University Press

Smith-Osborne, A. and Selby, A. (2010) Implications of the Literature on Equine-Assisted Activities for Use as a Complementary Intervention in Social Work Practice with Children and Adolescents. *Child & Adolescent Social Work Journal*, 27 (4): 291–307

Soren, I. (2001) *The Zen of Horseriding*. London: Little, Brown and Company

Southam-Gerow, M. and Kendall, P. (2002) Emotion Regulation and Understanding: Implications for Child Psychopathology and Therapy. *Clinical Psychology Review*, 22 (2): 189–222

Spradley, J. (1979) *The Ethnographic Interview*. New York: Holt, Rinehart & Winston

Staempflif, M. (2009) Reintroducing Adventure Into Children's Outdoor Play Environments. *Environment and Behaviour*, 41 (2): 268–280

Stein, M. (2006) Research Review: Young People Leaving Care. *Child and Family Social Work*, 11: 73–279

Strimple, E. (2003) A History of Prison Inmate-animal Interaction Programs. *The American Behavioural Scientist*, 47 (1): 70–78

Sustainable Development Commission (2008) *Health, Place and Nature: How Outdoor Environments Influence Health and Wellbeing: A Knowledge Base*, available online at: http://www.sd-commission.org.uk/publications.php?id= 712 [accessed on 17 January 2011]

Symington, A. (2012) Grief and Horses: Putting the Pieces Together. *Journal of Creativity in Mental Health*, 7 (2): 165–174

Taylor, A., Kuo, F. and Sullivan, W. (2001) Coping with Add: The Surprising Connection to Green Play Settings. *Environment and Behaviour*, 33 (1): 54–77

244 Bibliography

Teasdale, J., Williams, J., Soulsby, J., Segal, Z., Ridgeway, V. and Lau, M. (2000) Prevention of Relapse/Recurrence in Major Depression by Mindfulness-Based Cognitive Therapy. *Journal of Counselling and Clinical Psychology*, 68 (4): 615–623

Tedeschi, P., Fitchett, J. and Molidor, C. (2005) The Incorporation of Animal-Assisted Interventions in Social Work Education. *Journal of Family Social Work*, 9 (4): 59–77

Tester, K. (1991) *Animals and Society*. London: Routledge

The Consortium on the School-based Promotion of Social Competence (1996) The School-Based Promotion of Social Competence: Theory, Research, Practice, and Policy. In R. Haggerty, L. Sherrod, N. Garmezy and M. Rutter (Eds.), *Stress, Risk and Resilience in Children and Adolescents: Processes, Mechanisms, and Interventions*. Cambridge: Cambridge University Press

Thomas, G. and Thompson, J. (2004) *A Child's Place: Why Environment Matters to Children*. London: Green Alliance Demos, available online at: www.green-alliance.org.uk/uploadedFiles/.../A%20Childs%20Place% 20Final%20Version.pdf [accessed on 17 January 2011]

Thompson, R. (1994) Emotion Regulation: A Theme in Search of a Definition, *Monographs of the Society for Research in Child Development*, 59: 24–52 (2–3; Serial No. 240)

Thompson, K. and Gullone, E. (2008) Prosocial and Antisocial Behaviors in Adolescents: An Investigation into Associations with Attachment and Empathy. *Anthrozoos*, 21 (2): 123–137

Tottle, S. (1998) *Bodysense: Revolutionize your Riding with the Alexander Technique*. North Pomfret, Vermont: Trafalgar Square Books

Travers, M. (2001) *Qualitative Research Through Case Studies*. London: Sage

Trienbenbacher, S. (1998) The Relationship Between Attachment to Companion Animals and Self-Esteem. In C. Wilson and D. Turner (Eds.), *Companion Animals in Human Health*. London: Sage

Trotter, K., Chandler, C., Goodwin, D. and Casey, J. (2008) A Comparative Study of the Efficacy of Group Equine Assisted Counselling With At-Risk Children and Adolescents. *Journal of Creativity in Mental Health*, 3 (3): 254–284

Turner, W. (2005) The Role of Companion Animals Throughout the Family Life Cycle. *Journal of Family Social Work*, 9 (4): 11–21

Twigg, J. (2002) The Body in Social Policy: Mapping a Territory. *Journal of Social Policy*, 31 (3): 421–439

Ungar, M. (2004) The Importance of Parents and Other Caregivers to the Resilience of High-risk Adolescents. *Family Process*, 42 (1): 23–41

Ungar, M., Dumond, C. and Mcdonald, W. (2005) Risk, Resilience and Outdoor Programmes for At-risk Children. *Journal of Social Work*, 5: 319–338

Vidrine, M., Owen-Smith, P. and Faulkner, P. (2002) Equine-Facilitated Group Psychotherapy: Applications For Therapeutic Vaulting. *Issues in Mental Health Nursing*, 23 (6): 587–603

Voight, A. (1988) The Use of Ropes Courses as a Treatment Modality For Emotionally Disturbed Adolescents in Hospitals. *Therapeutic Recreation Journal*, 22 (2): 57–64

Waal de, F. (2009) *The Age of Empathy: Nature's Lessons for a Kinder Society*. New York: Harmony Books

Bibliography 245

Wagner, E., Rathus, J. and Miller, A. (2006) Mindfulness in Dialectical Behavior Therapy for Adolescents. In R. Baer (Ed.), *Mindfulness-based Treatment Approaches: Clinicians Guide to Evidence Base and Applications*. London: Academic Press

Walker, E. (2008) *Horse*. London: Reaktion Books

Walker, S. (2003) *Social Work and Child and Adolescent Mental Health*. Lyme Regis: Russell House

Walkerdine, V. (2008) Contextualizing Debates about Psychosocial Studies. *Psychoanalysis, Culture & Society*, 13: 341–345

Wals, A. (1994) Nobody Planted It, It Just Grew! Young Adolescents' Perceptions and Experiences of Nature in the Context of Urban Environmental Education. *Children's Environments*, 11 (3): 1–27

Walsh, F. (2009) Human-Animal Bonds II: The Role of Pets in Family Systems and Family Therapy. *Family Process*, 48 (4): 481–491

Ward Thompson, C., Tavlou, P. and Roe, J. (2006) *Free-Range Teenagers: The Role of Wild Adventure Spaces in Young People's Lives*. OPEN Space Research Centre Report for Natural England, available online at: www.countrysiderecreation. org.uk/Children%20Outdoors.pdf [accessed on 17 January 2011]

Webb, S. (2001) Some Considerations on the Validity of Evidence-based Practice in Social Work. *British Journal of Social Work*, 31 (1): 57–79

Wells, D. (2009) The Effects of Animals on Human Health and Well-Being. *Journal of Social Issues*, 65 (3): 523–543

Wells, N. (2000) At Home with Nature: Effects of 'Greenness' on Children's Cognitive Functioning. *Environment and Behaviour*, 32 (6): 775–795

Wells, N. and Evans, G. (2003) Nearby Nature: A Buffer of Life Stress among Rural Children. *Environment and Behaviour*, 35: 311–330

Werner, E (1993) Risk, Resilience, and Recovery: Perspectives From the Kauai Longitudinal Study. *Development and Psychopathology*, 5: 503–515

Werner, E. and Smith, R. (1992) *Overcoming the Odds: High Risk Children from Birth to Adulthood*. Ithaca, NY: Cornell University Press

WHO (2001) World Health Organization: International Classification of Functioning, Disability, and Health, available online at: http://www.who.int/ classifications/icf/en/ [accessed on 29 October 2010]

Widdicombe, S. (2008) *The BHS Book of the Natural Horse*. Cincinnati, OH: David & Charles

Wilkie, R. and Inglis, D. (Eds.) (2007) *Animals and Society: Critical Concepts in the Social Sciences Volume II: Social Science Perspectives on Human-Animal Interactions*. London: Routledge

Wilson, C. and Turner, D. (Eds.) (1998) *Companion Animals in Human Health*. London: Sage

Wilson, E. (1984) *Biophilia*. Cambridge, MA and London: Harvard University Press

Winnicott, D. (1953) Transitional Objects and Transitional Phenomena. *International Journal of Psychoanalysis*, 24: 88

Winnicott, D. (1965) *The Maturational Processes and the Facilitating Environment: Studies in the Theory of Emotional Development*. London: Hogarth Press

Winnicott, D. (1971) *Playing and Reality*. London: Tavistock

Winnicott, D. (1984) *Deprivation and Delinquency*. London: Tavistock

246   *Bibliography*

Wolkow, K. and Ferguson, H. (2001) Community Factors in the Development of Resiliency: Considerations and Future Directions. *Community Mental Health Journal*, 37 (6): 489–498

Yorke, J., Adams, C. and Coady, N. (2008) Therapeutic Value of Equine-Human Bonding in Recovery from Trauma. *Anthrozoos*, 21 (1): 17–30

Young, R. (2005) Horsemastership Part 1: Therapeutic Components and Link to Occupational Therapy. *International Journal of Therapy and Rehabilitation*, 12 (2): 78–83

Young, R. and Bracher, M. (2005) Horsemastership Part 2: Physical, Psychological, Educational and Social Benefits. *International Journal of Therapy and Rehabilitation*, 12 (3): 120–125

Zylowska, L., Ackerman, D., Yang, M., Futrell, J., Horton, N., Hale, T., Pataki, C. and Smalley, S. (2008) Mindfulness Meditation Training in Adults and Adolescents with ADHD. *Journal of Attention Disorders*, 11 (6): 737–746

Index

AAT (animal-assisted therapy)
  aims of recent research, 14
  American research, 14
  examples, 13–14
  historical perspective, 14
  overview, 13–14
  psychosocial benefits, 15
Achilles, 9
acts of helpfulness
  Minimax displays, 87–9
  reported benefits, 89
  resilience literature on, 47, 83,
    87–9, 145
Adams, C., 55, 56, 57, 69, 105, 106,
  108, 109, 126–7, 128, 134, 193
ADD (attention deficit disorder),
  impact of 'green' settings, 187
ADHD (attention deficit hyperactivity
  disorder), 30, 61, 64, 73, 84, 130,
  145, 157
  and being in nature, 175
adolescence
  appropriate types of therapeutic
    intervention, 32
  and risk, 43
adolescent delinquency, gendered
  perspective, 41
adolescent development, history of
  interest in, 40
adventure-based interventions, Moote
  and Wodarski's proposal, 32
Ainsworth, M., 53, 54
Aldridge, J., 29, 30, 49, 85, 179
Alexander Technique, 158
All, A., 4, 17, 19
Allen, N., 164, 167, 168
Almaas, A., 116
Anderson, R., 14
Anderson, W., 14
anger, reduction of, through
  therapeutic riding, 23

animal rights movement,
  beginnings, 10
animals
  differences between humans and,
    9–11
  exploration of humanity through, 9
  historical perspectives, 10
  prominence in children's literature,
    12–13
  religious perspectives, 10
  research and literature agenda, 12
  role as social lubricant, 13, 15
  social competence and contact
    with, 16
  and sociological perspective of
    children and, 12–13
  therapeutic importance to
    children, 13
  use of in research, 11
  value of in teaching responsibility
    to children, 13
animals and attachments, literature
  review, 107–8
anthropomorphism, 119, 129
antisocial behaviour
  AAT and, 16
  and lack of empathy, 113
Arluke, A., 8, 9, 12
Asperger syndrome
  and EAT/L, 181
  fun activities for young people with,
    73, 145
  see also autistic spectrum
asthma, 84
Atkinson, P., 63, 76, 79, 80, 197
at-risk, concept analysis, 35
at-risk young people, importance of
  alternative positive experiences
  for, 48–9
Atkinson, P., 76
Atlantis, 17

247

248 *Index*

attachment
  categories of insecure, 53
  Cinderella and Leo, 115–17
  Minimax and Ruby, 108
  trust and, 120
  types of, 53
attachment theory, 52–4
  criticisms, 56–7
  emergence, 38
  foundation, 52
  horses and relationships, 52
  human–animal relationship
    literature, 111–12
  literature review, 15, 107–8
  mindfulness and, 61, 158
  Scott on the importance of
    touch, 110
attention restoration theory, 31, 181
Atwood, E., 182, 183
Augustine, Saint, 10
authentic functioning, definition, 164
autistic spectrum, 73, 145
  benefits of EAT/L and TH, 157
  Lucy's diagnosis, 70
  Minimax, 132, 145
  and tactile stimulation, 110, 132
  value of animals for provision of
    tactile experiences, 132
  see also Asperger syndrome
Axline, V., 51, 57

Bachi, K., 4, 7, 19, 20, 23, 24, 33, 52,
  191, 194, 203, 205
Baer, R., 59
Baker, W., 16, 17, 25, 119
Baldwin, A., 204
Ball, D., 29, 30, 32, 93
Bandura, A., 46, 85
Bannister, A., 35, 37, 44, 84
Barclay, H., 4, 16, 17, 18, 25
bareback riding, 73, 90
Barker, S., 14
Barnardo's, 45
Barnes, R. K., 59, 155
Barr, R., 36, 38
Bass, M., 24
bear-baiting, 10
Beck, A., 14, 15, 45, 96, 191
Beck, L., 107, 111

Becker, B., 35, 50, 51, 96, 97
Becker, S., 29, 30, 49, 85, 179
behaviour
  AAT and, 15
  modifying through empathy,
    115–17
  of study participants, *see* TH sessions
being in nature
  benefits, 27, 30, 140, 154
  Nebbe on positive ways of being,
    32, 183
  study participants' feelings about,
    169–71
  see also the natural environment
Bekoff, M., 10, 12
Bell, S., 53, 54
Bellack and Bellack,
  psychoanalysts, 151
Bentham, J., 10
Bentley, J., 158
Berger, J., 25
Berger, R., 30–1, 154, 170, 176, 177
Bergin, A., 56, 190
Berman, D., 46, 48, 90
Bernstein, P., 14
Bertoti, D., 6, 19
Best, K., 2, 35, 36, 37, 39, 40, 41, 46,
  83, 84, 85, 97, 192
best practice agreements, 3
Bethel, 14
Bexson, T., 29
Biegel, G., 58, 59, 155, 157, 171,
  198, 200
biophilia theory, 3, 11, 16, 27–9, 177
bird song, 99, 110, 140, 166, 171, 174
Birke, L., 11, 25, 67
Birns, B., 56
Bizub, A., 7, 20, 22, 23, 24, 194
Blair, D., 32
Blake, H., 138, 161
Blaxter, L., 64, 77
blood pressure, and contact with
  animals, 12, 14
Bloor, M., 197
body awareness, 155, 158, 160
body language
  horses' impact on young peoples'
    awareness of, 155
  and horses' prey status, 155, 161–2

horses' understanding, 26–7
importance of learning about, 100
study participants gain
    understanding of and learn to
    modify, 73, 93–5, 127–8, 142–3,
    145–6, 150–1, 154–5, 157,
    159–60, 162–7, 169, 172
of wild animals, 172
body scan exercises, 59–60, 157,
    159–60, 162–3
body weight ratio, significance of
    similarities between
    mother-infant and
    horse-person, 57
Book of Job, 18
Boris, N., 52
Born, M., 37, 39, 40, 42, 43, 46, 91,
    92, 93
Bourdieu, C., 51
Bowers, M., 7, 20, 50, 52, 112, 122,
    126, 128, 137, 141, 194
Bowlby, J., 12, 38, 49, 52, 53, 54, 55,
    56, 57, 107, 111, 116
Boyce, T., 36, 38
boys, acceptability of animals as outlet
    to experience nurturing, 110–11
Bracha, S., 44
Bracher, M., 6, 23
Bradberry, J., 52, 105, 126
Braff, E., 46, 90
Brandt, K., 25
Braun, V., 79
Bretherton, I., 53, 117
Bright Roberts, F., 52, 105
Brock, B., 19
Broersma, P., 6, 17, 25, 194
Bronfenbrenner, U., 36, 37, 40,
    84, 201
Brooks, S., 7, 15, 20, 54, 55, 110, 126,
    130, 132, 136, 137, 146, 172, 193,
    198, 199
Brown, K., 59, 61, 155, 157, 161,
    164, 168
Brownlee, K., 57
brutal practices, used towards
    animals, 10
Buddhism, 58, 116, 154
Budiansky, S., 25
Budiansky, S., 16, 25, 161

Burdette, H., 29, 32
Burgon, H., 20, 27, 50, 127, 194
Butler, A., 197

Cahalan, W., 179
calming effect, of the horses, 132–3,
    137, 156–8, 171–2
calmness, animals' requirements, 172
Cannon, W. B., 44
care, children in, *see* children in care
care leavers
    entering higher education, 88
    GCSE performance, 46
caring for horses, daily activities, 184
caring for pets
    and empathy, 112
    gendered perspective, 110–11
Carns, A., 32, 46, 48, 49, 90, 200
Carns, M., 32, 46, 48, 49, 90, 200
Castellan, B., 57
cave paintings, horses in, 16–17
Cawley, M., 17, 50, 194
Cawley, R., 17, 50, 194
CBEIP (Certification Board for Equine
    Interaction Professional), 2
Celts, 17
cerebral palsy, research on the benefits
    of hippotherapy for children
    with, 19
challenge, motivational power, 100–2
Chamberlin, J., 8, 16, 17, 18, 25
Chambers, R., 164, 167, 168
Chandler, C., 11, 13, 14, 15, 29
Chardonnens, E., 7, 20, 69, 121,
    128, 129
Chevalier, V., 37, 39, 40, 42, 46, 91
child-abuse
    'Baby P' case, 44
    neurobiological perspectives, 44–5
    socioeconomic response, 45
child-animal relationship, lack of
    scholarly interest, 12
childhood development
    impact of alienation from nature
        on, 28
    impact of multiple psychosocial risk
        factors on, 36
    impact of parenting, 52–3
    importance of empathy for, 112

250 *Index*

childhood obesity, 29–30, 187
childhood risks, and negative outcomes, 35
childhood traumas, embodiment of, 132
children
 sociological perspective of animals and, 12–13
 therapeutic importance of animals to, 13
 Western cultural traditions, 12
children in care
 academic achievement, 87–8
 similarity between horses and, 138
 see also care leavers
children's literature, prominence of animals in, 12–13
chimeras, 11
Chiron, 17
Chodorow, N., 56
Christensen, P., 12, 51, 81
Christianity, legitimisation of human animal difference, 10
Cicchetti, D., 35, 36, 39, 46, 50, 51, 83, 85, 91, 96, 97, 124, 201
Cinderella, 73, 78, 100, 113–18, 122–3, 133, 141, 143, 152, 156, 161, 164–70, 182, 185, 187, 206, 220
Clarke, S., 37, 39, 62, 63, 78, 197
Clarke, V., 79
Clarkson, P., 68, 150
Cleary, R., 57
Coady, N., 56, 57, 69, 105, 106, 108, 109, 126, 127, 128, 134, 193
cock-fighting, 10
Coffey, A., 63, 76, 79, 80, 197
cognitive-behavioural approach
 and the normative model, 50–1
 to resilience and coping, 45
 vs socio-economic approach, 40
cognitive-behavioural therapy
 criticisms, 45
 emergence, 45
cognitive functioning, and engagement with the natural environment, 32, 177
Coleman, M., 59, 154, 163, 168
communication, mobile methods, 91

community-based interventions
 benefits for at-risk young people, 49–50
 traditional, 49
community groups, importance of, 97
community participation, importance for health and functioning, 206
companionship, pets' provision, 15
Compas, B., 39, 40, 41, 42, 43, 92
confidants
 horses as, 121–4
 literature on animals as, 121–2
coping, and resilience, 38–9
Corrigan, P., 75
Corson, E., 13, 15, 107
Corson, S., 13, 15, 107
Crane, L., 4, 17, 19
Crawford, E., 15, 112
Creswell, D., 61, 161, 164
Cushing, J., 20, 24, 27
cruelty to animals, historical acceptability, 10

Daly, B., 15
Daniel-McKeigue, C., 190
Davidson, L., 7, 22, 23, 24, 194
Davies, J. A., 19
Davies, P., 124
Davies, S., 8, 17
Dawson, K., 14
Day, P., 76
D'Cruz, H., 78, 196, 200
Delamont, S., 63, 76, 80, 197
delinquency, gendered perspective, 41
Dell, C., 204
Denscombe, M., 77, 197
Denzin, N., 62, 81, 196, 200
De Paul, J., 105, 112, 117
depression, 37, 40, 42, 51, 84, 89, 112
Descartes, R., 9
Desmond, B., 67
Devall, B., 11, 31
diet and exercise, learning about the importance of, 185–6
directive work, ethical perspective, 149
dolphins, 13
Dominelli, L., 51
Donaghy, G., 29

donkey-facilitated therapy, 13–14
Dorrance, B., 67
Dossey, L., 11, 13, 14, 15
dream symbolism, horses' role, 182
dressage, 17, 122, 144
Dryden, J., 36, 37, 38, 39, 40, 46, 47,
    48, 50, 83, 84, 89
Duchowny, C., 24
Dumond, C., 32, 48, 49, 64, 84, 90,
    98, 200
Dunayer, J., 11
Durr, P., 29

EAGALA (Equine-Assisted Growth and
    Learning Association), 2
EAP (equine-assisted psychotherapy)
    Esbjourn's research, 133, 136
    evaluation studies, 24
    international recognition, 24
    trauma victims study, 134
    see also EFP
EAT/L, term analysis, 2
EAT/L and TH
    aims, 95
    counter-indications, 204–5
    definition, 207
    evolution, 2
    lack of research, 4
    reason for effectiveness in
        therapeutic work, 25–7
    see also EAT/L studies; TH sessions
EAT/L studies
    challenges for, 24
    disparate paths, 33
    EFP programme, 23–4
    literature review, 19–24
    therapeutic horse-riding programme
        for adults, 22–3
    vaulting study, 21–2
    wild mustang programme, 20–1
Ebberling, C., 29, 30, 187
ecotherapy, 29, 33, 169
eczema, 84
Edney, A. T., 16
EFP (equine-facilitated psychotherapy)
    Hayden's findings, 98
    research needs, 22
EGEA (Equine Guided Education
    Association), 2

Egeland, B., 37, 84
Ellis, A., 45
Ellis, J., 46, 90
Emma, 70–1, 84, 91, 93, 100–1, 113,
    118, 123, 133, 135–6, 143–5,
    151–2, 155, 172–3, 178, 181–2,
    185, 220
emotional distress, horses' innate
    sensitivity to, 105
Emotional Intelligence (Goleman), 102
emotional security theory, Davies,
    Winter and Cicchetti's claim, 124
emotion regulation
    awareness of body language
        and, 172
    mindfulness and, 164
    and natural places, 175–6
empathetic parenting, 112
empathy
    childhood pet-keeping and, 112
    horses and the development of,
        111–14
    importance for child
        development, 112
    lack of and antisocial behaviour, 113
    Lexman and Reeves' description,
        112–13
    modifying behaviour through,
        115–17
    necessity for facilitation of
        therapeutic change, 128
    participants' ability to display with
        horses, 105
    as precursor to nurturance, 112
empowerment
    and identification, 91–3
    inherent limitations and
        contradictions, 81
endorphins, 184
engagement, eye contact and, 91
Engel, B., 6
Epona, 2
equine-assisted therapy/learning, *see*
    EAT/L
Erikson, E., 41, 43, 92
Esbjourn, R., 7, 133, 136
Evans, G., 31
Evans, R., 58, 122
Ewert, A., 48, 84, 200

252 Index

Ewing, C., 7, 24, 26, 33, 50, 137, 141, 146, 198
exercise, and health, 29–30
experiences, of study participants, *see* TH sessions
experiential therapy, 21
exploitation of animals, historical perspective, 10–11
eye contact, and engagement, 91

Farrell, A., 81
Faulkner, P., 7, 20, 21, 22, 24, 33, 50, 52, 55, 57, 102, 126, 128, 136, 141, 143, 147, 194
fear, overcoming through EAT/L, 23
Fearnley, S., 111, 116
The Federation of Horses in Education and Therapy International (HETI), 3
felt security, 111
Ferguson, H., 39, 40, 48, 49
Fine, A., 11, 12, 13, 14, 15, 26, 56, 94, 106, 128, 133, 135, 137
Finlay, L., 62, 63
Fitchett, J., 13, 15, 107, 111
Foard, N., 63, 75
Folkman, S., 36, 38, 45
Fook, J., 51, 200
Ford, D., 197
Forrester, D., 205
Foster, A., 45
foster care
    Cinderella's story, 114, 117
    difficulty of tactile contact, 108, 131
    impact of moving between numerous placements, 46
    Kelly's story, 69, 139, 152
    Minimax's story, 87
    system analysis, 64
    Wayne's story, 148, 150
Frame, D., 7, 54, 55, 126, 138, 192, 193, 198, 205
France, cave paintings, 17
Frank, J. B., 55, 109, 125, 133, 149, 150
Frank, J. D., 55, 109, 125, 133, 149, 150
Franklin, A., 8, 10, 12, 58, 122
Frantz, D., 183

freedom, horses' representation, 182
Freeman, L., 14
Freud, S., 138
Frewin, K., 16, 18, 24, 33
Freya, 71, 93–5, 121, 133, 158–61, 163, 168, 176–7, 181, 220
Friedmann, E., 4, 11, 12, 14, 123, 199
Friesen, L., 126, 133, 137, 204
Fuller, R., 62, 63
fully functioning person, 60
Fulton, P., 58
future research, suggestions for, 205–6

Gaia hypothesis, 28, 179, 189
Game, A., 16, 58, 122, 155
Gammage, D., 52, 55, 56, 57, 68, 112, 117
Garcia, D., 4, 7
gardening project, *see* therapeutic horticulture
Gardiner, B., 16, 18, 24, 33
Garfield, S., 190
Garmezy, N., 2, 35, 36, 37, 38, 39, 40, 41, 46, 83, 84, 85, 97, 192
Garner, L., 45
Garrett, P., 199
Geertz, C., 62
Gehrke, E., 27, 33, 123, 161, 204
Gerhardt, C., 39, 40, 41, 42, 43, 44, 92
Gerhardt, S., 113
Germany, use of AAT, 14
Germer, C., 58, 59, 60, 155, 157
gestalt approaches, 62, 68, 140, 157, 179
    see also in the present moment; mindfulness
Gill, T., 29, 30, 32, 93
Giller, H., 98
Gilligan, R., 39, 43, 47, 48, 50, 92, 97
Glad, K., 44
Glantz, M., 37
Goldwyn, R., 52
Goleman, D., 102, 192
Gomes, M., 10, 11, 27–8, 31, 170, 200
good enough parenting, 55, 111, 136
Goodman, P., 68, 140
Goodman, T., 58, 59
Gordon, D., 27, 138
Gough, B., 62, 63

Index 253

GP visits, pet ownership and, 15
Graham, J., 205
Graue, E., 51
Gray, J. A., 44
Greek mythology, horses in, 9, 17
Green, J., 52
Greene, S., 81
'green' exercise, 29–30
Greiger, R., 45
grooming, 90, 109, 131–2, 141, 148, 158–9, 184, 207
Grute, J., 28, 29, 138, 139, 140
Gubrium, J., 196, 197
Guibert, M., 105, 112, 117
guide dogs, 13
Gullone, E., 105, 112, 164, 167, 168
Gunner, M., 36, 44, 84

hacking, 114, 123, 176
Hagell, A., 98
Haggerty, R., 37, 38, 40, 51
Hahn, H., 37, 39
Hallberg, L., 16
Halliwell, E., 15, 27, 29, 36, 49, 58, 59, 155, 156, 184
Hammen, C., 41
Hampel, P., 39, 40, 41, 43, 92
Hannah, B., 90, 183
Haraway, D., 10, 11, 12
Hargreaves, I., 58, 59
Hart, B., 14
Hart, L., 19, 107
Hayden, A., 7, 84, 98, 205
healing, and encounters with horses, 119
healing power of nature, literature review, 170
health
    exercise and, 29–30
    importance of community participation for, 206
healthy touch, importance of, 55
heart attack, survival rates of pet owners, 14
heart rates, research into impact of being with horses on, 122–3
Heerwagen, J., 176
Hefferline, R., 68, 140
Heide, K., 204

Held, C., 183, 205
helpfulness, acts of, *see* acts of helpfulness
'helping others helps oneself', Schwartz and Sendor's claim, 89, 112
Henn, M., 63, 75
Henrickson, J., 6
Heppner, W., 59, 61, 161, 164, 166, 167
herd structure, understanding, 25
Hertz, R., 63, 80
high identification, Jackson, Born and Jacob's claim, 92
high-risk peer groups, alternative provision, 49
Hinden, B., 39, 40, 41, 42, 43, 92
Hine, R., 15, 29, 30
Hingley-Jones, H., 76
Hippocrates, 17
hippotherapy
    research base, 6
    research on the benefits of, 19
Hogan, D., 81
Hoggett, P., 37, 39, 45, 51, 62, 63, 197
holding environment
    concept analysis, 54
    and EAT/L, 54–6
    how horses can help, 55
    literature review, 136
    primary aspects, 54
Holland, J., 32, 46, 48, 49, 81, 90, 200
Holland, S., 203
Hollway, W., 44, 62, 63, 77, 78, 196
Holmes, J., 56
Holstein, J., 196, 197
Hooker, S., 14
hoping, Brooks on, 110
the horse
    cross-cultural reverence towards, 17
    etymology, 17
    inherent vulnerability, 25
The Horse Boy (Isaacson), 1
horse management, and the natural environment, 172–5
horse-related sports, and the human need to take risks, 18

254   *Index*

horses
accessibility of mindful experiences, 168
benefits of positive experiences with, 98
calming effect, 132–3, 137, 156–8, 171–2
as a challenge, 100–2
and the client–therapist relationship, 125
as confidants, 121–4
cultural perspective, 17–18, 182
and the development of empathy, 111–14
effectiveness as therapists, 126
emotion regulation through experiences with, 102
horses, myth and magic, 182–3
horses, nature and relationships, 177–9
importance of their sociological and cultural role, 25
innate sensitivity to emotional distress, 105
in Jungian dream interpretation, 182
learning and taking risks outdoors with, 31–3
link between nonjudgmental otherness and feeling safe, 147
mastery of and self-esteem, 85–7
mindfulness and the calming influence of, 156–8
mindfulness of, 161–4
and the natural environment, 27
necessity of self-awareness for successful partnership with, 164
occupation of space 'in between' other animals, 25
and present-day therapy, 17
prey status, 55, 155, 158, 161
psychosocial and psychospiritual dimension, 168–9
reason for effectiveness in therapeutic work, 25–7
relationship of humans and, 16
requirement for a successful partnership, 155
and a safe space, 132–6
sensitivity to aggressive or agitated behaviour, 164
similarity between children in care and, 138
study participants, 71–2
and the therapeutic relationship, 57–8
unexplained actions and encounters between people and, 119
as vehicle for projection of a participant's unconscious worries and fears, 149–50
ways of helping in the holding environment, 55
weather's influence, 32
working outside with, 186–7
horticulture
social and therapeutic, 30
see also therapeutic horticulture
Houpt, K., 25
Howard, S., 36, 37, 38, 39, 40, 45, 46, 47, 48, 50, 83, 84, 89, 91, 97
Howe, D., 52, 111, 116, 198, 199
Howey, M., 8, 9, 18, 182
Hughes, C., 64, 77
Hughes, D., 35, 37, 52, 83, 84, 109, 111, 116
human–horse communication, 25
Humblet, I., 37, 39, 40, 42, 46, 91
Humphries, B., 81
hunting, and horse related sports, 18
Hutchinson, S., 46, 90
Hyland, A., 16, 25

identification
empowerment and, 91–3
as protective factor, 92–3
Illich, I., 51
immune system, stress and the, 37, 84
in the present moment, 58–9, 130, 141, 155, 160
incarceration, gendered perspective of outcomes, 42
individual appraisal, as significant factor in resiliency, 38
infant behaviour, and later behavioural problems, 42
Inglis, D., 8, 10
insecure attachment, categories, 53

Index   255

insurance cover, 73, 147
interventions, traditional
    community-based, 49
interviewing, psychosocial
    approach, 78
introjections, 125, 152
invisible riding, 59–60, 73, 89–91,
    157, 158–61
Irvine, L., 12
Irwin, C., 26
Isaacson, R., 1, 17, 33, 105, 119,
    122, 128

Jack, G., 176
Jackson, S., 8, 18, 35, 37, 41, 43, 46,
    64, 87, 88, 92, 182, 183
Jacob, M-N., 43, 92, 93
James, A., 12, 51, 81
Jefferson, T., 62, 63, 78, 196
Jennings, G., 14
Johnson, B., 36, 37, 38, 39, 40, 45, 46,
    47, 48, 50, 83, 84, 89, 91, 97
Johnson, J., 37
joining-up
    definition, 94
    Freya tries, 93–5
    function of exercise, 73
Jones, B., 18
Jones, M., 78, 196, 200
Jones, N. A., 8, 17
Jones, P., 97, 149
Jones, R., 9, 183
Joy, A., 7, 22, 23, 24, 94, 194
Jung, C. G., 9, 21, 125, 138, 182, 183
Jungian perspectives
    horses in dream interpretation, 182
    the natural environment, 169
    role of fairy tales in child
        development, 183
    the therapeutic relationship, 125

Kabat-Zinn, J., 58, 59
Kahn, P., 170
Kaiser, L., 7, 19, 20, 23, 24, 33, 50, 194
Kanner, A., 10, 11, 27, 31, 170, 200
Kaplan, S., 31, 169, 170, 176, 181
Karen, R., 49, 52, 53, 112
Karol, J., 7, 20, 25, 26, 52, 54, 102,
    107, 112, 126, 143, 193

Katcher, A., 14, 15, 16, 28, 107, 177
Katz, M., 46, 47, 85, 89, 90, 97
Kauai, life-span study, 42
Kellert, S., 11, 27, 28, 29, 170, 176, 177
Kellett, M., 51
Kelly, 49, 69, 71, 80, 84, 89–91, 98–9,
    121, 127–9, 133–5, 137, 139, 152,
    179, 183, 220
Kemp, K., 19, 20, 24, 33
Kemp, M., 9, 10
Kendall, P., 102, 103, 192, 193
Kernis, M., 61, 161, 164, 166
Kiley-Worthington, M., 161
Kimball, C., 194
Klein, M., 41
Klontz, B., 7, 20, 23, 26, 27, 33, 136,
    143, 194
Knox, J., 9
Knox, M., 40, 50, 57, 136, 146, 205
Kogan, L., 15, 87, 89
Kohanov, L., 2, 6, 16, 25, 27, 105,
    119, 128
Korpela, K., 29, 30, 31, 176, 181
Kraemer, J., 39
Krane, J., 57
Kunzle, U., 6
Kuo, F., 29, 30, 49, 175, 187
Kurtz, R., 132, 133
Kuyken, W., 59

Laing, R. D., 50
Laird, J., 62
Lambert, M., 56
language, Dunayer on speciesism
    in, 11
Latimer, J., 25, 67
Lawrence, E., 18, 25
Lazarus, R., 36, 37, 38, 45, 46, 85, 96
leading, 73, 93, 118, 142, 148
Lentini, J., 40, 50, 57, 136, 146, 205
Leupnitz, D., 56
Levi-Strauss, C., 8
Levine, P., 36, 37, 84
Levinson, B., 11, 13, 14, 15, 16, 28,
    106, 107, 109, 110, 111, 113, 121,
    126, 171, 199
Levinson, D., 38
Levi-Strauss, C., 8
Lexman, J., 105, 112, 113, 192

256 *Index*

life-span, gendered perspective of protective factors, 42
Lincoln, Y., 62, 81
Linden, S., 28, 29, 138, 139, 140
Llabre, M., 24
llamas, 172
Llewellyn, A., 33
Loch, S., 17
Louv, R., 10, 11, 28, 29, 49, 170, 175
Lovelock, J., 28, 170, 179, 189
Loving, G., 4, 17, 19
Lucy, 40, 69–72, 74, 78–9, 91–3, 95–6, 98, 100, 114, 123–4, 130, 142, 156, 176–8, 180–3, 220
Ludwig, D., 29, 30, 187
Luther, S., 35, 50, 51, 96, 97
Lynch, R., 199
Lyons-Ruth, K., 44

MacDonald, P., 7, 20, 50, 52, 112, 122, 126, 128, 137, 141, 194
Mackewn, J., 68
MacKinnon, J., 6, 19
Madresh, E., 15, 107, 111, 191
magical powers, early man's belief in the horse's, 17
Malaspina, A., 14
Mallon, G., 14, 29
March, J., 59
Marr, C., 14
Martinisi, V., 25
Maslow, A., 51, 57
Mason, O., 58, 59
Masten, A., 2, 35, 36, 37, 39, 40, 41, 45, 46, 83, 84, 85, 97, 192
mastery, impact on self-esteem and self-confidence, 85–7
mastery and control, as resilience enabling characteristics, 46
maternal deprivation, impact of, 52–3
Matsukawa, J., 44
May, T., 63, 79, 197
Mayberry, R., 17
Mazis, G., 11
McCartney, K., 38
McCormick, A., 6, 16, 17, 25, 26, 27, 52, 55, 57, 102, 105, 112, 126, 128, 143
McCormick, B., 48, 200

McCormick, M., 6, 16, 17, 25, 26, 27, 52, 55, 57, 102, 105, 112, 126, 128, 143
McCormick, T., 6, 16, 17, 18, 25
McCurdy, L., 29, 187
Mcdonald, W., 32, 48, 49, 64, 84, 90, 98, 200
McGreevy, P., 25
McLeod, J., 30, 31, 75, 154, 170, 176, 177
McLoughlin, B., 149
McParlin, P., 35, 37, 41, 46, 64, 87, 88
McVeigh, T., 45
Mearns, D., 51, 56, 116, 125, 152
medical science, animals and, 11
meditation techniques, 60, 163
Meinersmann, K., 52, 105, 126
Melson, G., 8, 12, 13, 14, 15, 106, 107, 110, 111, 112, 113, 117, 151, 152, 199
mental health
and the development of self-confidence and self-esteem, 85
and 'green' exercise, 30
therapeutic horse riding programme, 22
Merz-Perez, L., 204
Miller, A., 58, 59
mindfulness
accessibility of mindful experiences with horses, 168
attachment theory and, 61, 158
and authentic functioning, 61, 164–8
and body language, 161–2
and the calming influence of horses, 156–8
child psychotherapy, TH and, 154–6
Cinderella and Duchess, 164–7
concern with *being*, 60
definition, 58, 155
emotion regulation and, 164
and the expression of negative thoughts and feelings, 62
of horses, 161–4
introduction into mainstream health services, 59
'invisible' riding and, 158–61

and links to therapeutic
horsemanship, 58–62
opposite of, 155
relevance to children, 58–9
roots, 154
techniques, 59
Minimax, 69, 72–3, 75–6, 82, 86–9,
91–2, 96–7, 100, 106, 108–9,
118–21, 129–33, 138, 141–2,
145–7, 152, 156, 170, 173–5, 181,
184, 186–7, 220
mirroring, 26–7, 162
Misra, M., 29
Mistral, K., 26, 27, 123
Moffitt, T., 41
Molidor, C., 13, 15, 107, 111
moment-to-moment
concentration, 158
see also gestalt approaches
Mongolian Shamans, 17
Moorhead, J., 122
Moote, G., 32, 48, 49, 64, 200
Moreau, L., 52
Morgan, P., 32
Morgan, S., 163, 164
Morgan, W., 163, 164
Morrow, V., 51
Morton, L., 15, 16
Moss, D., 59, 155
Moustakas, C., 63
Muñoz, S., 15, 29, 30, 32, 49, 187
Murray, J., 13
Murray, R., 49
Myers, G., 12, 15, 47, 89, 106, 113

Narey, Martin, 45
the natural environment
cognitive functioning and
engagement with, 32, 177
connection and listening to your
instinct, 179–80
freedom, escape and getting away
from problems, 180–2
Gaia hypothesis, 179
horse management and, 172–5
horses, myth and magic, 182–3
horses and relationships, 177–9
importance of in TH, 170–2
learning opportunities and, 175–7

literature on the need of for
help, 170
and natural horsemanship, 172
physical activity and learning in,
183–6
psychosocial and psychospiritual
dimension of horses and, 168–9
research exploring the link between
health and, 49
restorative character, 170–2
space to let off steam, 186–7
as therapeutic space, 30–1
see also being in nature
natural horsemanship
definition, 67
system and philosophy, 172
and unconditional positive
regard, 129
Nebbe, L., 27, 32, 170, 179, 183
Neckoway, R., 57
negative life outcomes
attributed to risk factors in
childhood, 37
usefulness of therapeutic
interventions to avoid, 84
Netting, F., 14, 107
Nettles, S., 39
neurobiological approach
criticisms, 44
to risk and resilience, 44–5
New, J., 14, 107
New Mexico, wild mustang therapy
programme, 20–1
NGC (Natural Growth Centre), 138–40
Nightingale, F., 13
non-directive approaches, 149
nonjudgmental otherness, and
feelings of safety, 147
nonjudgmental relationships,
importance of, 126
North American Indians, 17
nurturing, pets as acceptable outlet for
boys to experience, 110–11

obesity, childhood, 29–30, 187
object presentation, 54–5, 136
Obradovic, J., 36, 37, 40, 45
O'Brien, L., 49

258 *Index*

offender rehabilitation, American use of equine therapy-based programmes, 20–1
O'Hara, M., 63
Okely, J., 75
Orians, G., 176
O'Rourke, K., 204
outdoor experiences, and protective mechanisms, 98
outdoor pastimes
consequences of reduction in, 28–9
importance for both children and adults, 49
outdoors, learning and taking risks with horses, 31–3
outward bound courses, 32
overcoming fear
benefits, 93, 103
Freya and Hector, 93–5
Lucy and Timmy, 114
riding and, 23
Wayne's experience, 97
Owen-Smith, P., 7, 20, 21, 22, 24, 33, 52, 55, 57, 102, 126, 128, 136, 141, 143, 147, 194

parenting, development impact, 52–3
Parish-Plass, N., 15, 54, 107, 111
PATH Intl (Professional Association of Therapeutic Horsemanship International), 2
Patton, M., 77
Paul, E., 11, 12, 14, 15, 16, 112
Pawlak, D., 29, 30, 187
Payne, G., 63, 196
Peacock, J., 15, 29, 30, 154
Pech-Merle, 17
peer-group identification, 92
Pegasus, 17
Pemberton, C., 108, 131, 199
Perls, F., 68, 140
pet-assisted therapy, expansion in the UK, 13
Petch, A., 62, 63
Petermann, F., 39, 40, 41, 43, 92
pet-keeping, Melson on, 12
pet ownership
impact on health and well-being, 15
in the UK, 13

pets
as acceptable outlet for boys to experience nurturing, 110–11
psychosocial benefits, 14
Philo, C., 8, 9, 10
physical activity, and educational engagement, 169
physical touch, difficulties and importance of, 108–9, 131
Place, M., 35, 37, 47, 49, 85, 96, 97, 201
Pleck, J., 39
Podberscek, A., 11, 12
Poresky, R. H., 13, 15, 113
positive opportunities
importance of exposure to, 97
opening up, 97–100
Rutter's term, 97
transferable skills and, 95–7
practitioner requirements, 3
present moment, 58–9, 130, 155
Pretty, J., 15, 29, 30
prey status, of horses, 55, 155, 158, 161
Price, J., 44
Priest, P., 29
prisons
benefits of equine therapy, 20–1
impact of AAT provision in, 16
professional organisations, 2–3
projection, 27, 62, 78, 116, 118, 125, 138, 145, 149
protective factors
definition, 39
gendered perspective, 42
identification, 92–3
and the importance of alternative positive experiences, 48
incorporated by community-based interventions, 49
mastery and, 46
peer group identification, 43, 92–3
protective processes vs, 48
and risk and resilience mechanisms, 39–40
in successful parenting and child development, 112
usefulness of EAT/L and TH in providing, 84, 97

protective false self, 54, 56, 116
Prout, A., 12
psychosocial approach, vs
    neurobiological, 44
psychosocial interviewing,
    techniques, 78
psychosocial risk factors, 36
psychotherapy, 15, 55–6, 58, 64, 75,
    78, 123, 132, 138, 146
  animal facilitation, 15
  most powerful predictor of positive
    client outcome, 56
Pugsley, L., 80, 81

Quaker movement, 14
Quevedo, K., 36, 44, 84

ragwort, 174
Rainer, P., 15
Rallis, S., 76
Ralston, T., 44
Rashid, M., 25, 26, 67, 138
Rathus, J., 58, 59
RDA (Riding for the Disabled),
    formation, 19
recidivism, AAT and, 16
Rector, B., 27, 146
Rees, L., 25, 26, 67, 68, 138, 158, 161
Reeves, R., 105, 112, 113, 192
reflexive approach, power issues, 80–2
Regan, T., 12
Reid, C., 14
relationship, therapeutic, see
    therapeutic relationship
relationship-building
  Cinderella and Leo, 113–14
  and confidence, 89–91
  Kelly and Ruby, 127–8
  Minimax and Jason, 145–7
  Minimax and Ruby, 108
  Minnie helps Lucy with, 177–8
  through unconditional positive
    regard, 127–9
relationships, horses, nature and,
    177–9
Remick-Barlow, G., 7, 20, 24, 33, 56
research
  overview of existing, 6–7
  use of animals in, 11

research project
  aims, 207
  child protection and disclosure
    policy and procedure for
    therapeutic horsemanship,
    213–14
  consent forms, 209–10, 212
  consent process and ethical issues,
    73–4
  data analysis procedures, 79–80
  data-collection procedures and
    instruments, 75–80
  equine participants, 71–2
  fieldnotes, 76
  fieldwork, see TH sessions
  horse management policy and code
    of ethics, 215
  human participants, 69–71
  implications for social work practice
    and related fields, 197–200
  information sheets, 207–8, 211
  insights from, 191–5
  interviews, 77–8
  introduction, 218
  limitations, 195–7
  methodological and theoretical
    contributions, 201–4
  methodology, 75
  natural horsemanship approach,
    67–9
  participants, 207, 220–1; see also
    Cinderella; Emma; Freya; Kelly;
    Lucy; Minimax; Wayne
  PhD questions for young people,
    218–19
  philosophy and practice, 68–9
  problems and power issues, 80–2
  pseudonyms, 74–5
  questionnaires, 78–9
  reflections on resilience as related
    to, 200–1
  setting and location, 64–6
  site and methods, 62–4
  TH sessions, 72–3; see also TH
    sessions
  therapeutic horsemanship
    questionnaires, 216–17
  types of activities, 73

260 Index

resilience
  connection of protective
    mechanisms to, 98
  coping and, 38–9
  definition, 35, 83
  enabling characteristics, 45
  gendered perspective, 42
  mastery and control as enabling
    characteristics, 46
  qualities of identifiable from
    infancy, 42
  risk and, *see* risk and resilience
  Rutter's warning, 48
  variables associated with, 39
resilience theory, criticisms, 50–2
responses, of study participants, *see*
  TH sessions
responsibility
  impact on self-esteem and
    self-efficacy, 89
  value of animals in teaching
    children, 13
restorative environment, 170
Retter, K., 17, 50, 194
Rew, L., 107, 113
rhythm of riding, therapeutic
  quality, 122
rhythmic harmonisation, horse's gait
  and, 57–8
Richards, M., 51
Richards, S., 119
riding
  and empowerment, 91, 97, 99, 183
  and overcoming fear, 23
  and physical contact, 55
  requirements for, 155
  sensorimotor experience, 19
  therapeutic effects of the
    rhythm, 122
  therapeutic history, 16–17, 19
  see also bareback riding; invisible
    riding
Riessman, C., 63
risk
  benefits of for self-development, 93
  development model, 40
  indicators of, 51
  learning and taking risks outdoors
    with horses, 31–3

risk and resilience
  in adolescence, 43
  behavioural and developmental
    perspectives, 36
  cognitive-behavioural approach,
    45–6
  community protective factors and
    EAT/L, 48, 50
  coping and resilience, 38–9
  and coping strategy
    development, 96
  development model, 40–1
  gender differences, 41–3
  importance of self-confidence and
    self-esteem for, 85
  literature review, 35–7
  neurobiological approach, 44–5
  parameters of research, 84–5
  protective factors and mechanisms,
    39–40
  required acts of helpfulness, 47–8
  resilience theory criticisms, 50–2
  understanding risk factors, 36–8
risk factors, 35–8, 46–7, 50, 83–5, 102
risk-taking, 18, 32, 49, 159, 183, 187
Risley-Curtiss, C., 15
Robbins, L., 7, 20, 24, 33, 56
Roberts, M., 25, 26, 67
Roberts, J., 25
Robinson, C., 51
rodeo, Lawrence's study, 18
Roe, J., 187
Rogers, C., 50, 51, 56, 57, 68, 69, 116,
  125, 126, 128
Rogosch, F., 35, 36, 39, 46, 83, 85, 91,
  96, 97, 201
Rolfe, J., 158
Rosenthal, S. R., 18, 93
Ross, N., 91
Rossman, G., 76
Roszak, T., 10, 11, 27, 3, 170, 200
Rothbaum, F., 204
Rothe, E., 7, 24, 146
Runnquist, A., 8
Rutter, M., 2, 35, 36, 38, 39, 46, 47,
  48, 49, 56, 83, 93, 97, 98, 192
Ryan, R., 59, 61, 155, 157, 161,
  164, 168
Rycroft, C., 149

Sable, P., 15
sacred spaces, 31, 176
safety, horses and a feeling of, 132–6
Saunders, C., 8, 9
Scanlan, L., 5, 18, 183
Scarr, S., 38
Schiltz, P., 204
Schultz, P., 7, 24, 33, 56
Schwartz, C., 89, 112
Scott, J., 110, 132, 199
Sedgewick, D., 123, 133
Segal, Z., 59, 60, 62, 157, 160, 163, 167, 168
Selby, A., 4, 20, 40, 84, 190, 203, 205
self-awareness
    adolescent levels, 168
    importance of for an effective therapist, 125–6
    necessity for successful partnership with horses, 164
self-confidence
    mental health and the development of, 85
    risk and, 93
self-efficacy
    and the concept of required helpfulness, 47
    and control of behaviour, 96
    definition, 85
    impact of responsibility on, 89
    increasing through mastery experiences, 46
    overcoming fear and, 95
    promoting through outdoor experiences, 98
    as resilience enabling characteristic, 45
    risk and, 93
    and self-esteem, 85–6
self-esteem
    AAT and, 15
    community interventions and, 50
    Ewing's findings, 24
    gendered perspective, 42
    the horse's usefulness in building, 20
    identification and, 92

impact of being left behind developmentally and educationally, 46
impact of responsibility on, 89
impact on development, 43
improvement in and attitudes towards others, 87
mastery of horses and, 85–7
mental health and, 85
promoting through outdoor experiences, 98
as resilience enabling characteristic, 45
responsibility of looking after an animal and, 16
risk and, 93
secure vs fragile, 61
self-efficacy and, 85–6
self-harming, 37, 84
self-image, as resilience factor, 42
Sempik, J., 29, 30, 49, 85, 179
Sendor, R., 89, 112
Serpell, J., 8, 11, 12, 13, 14, 15, 16, 112
Sessions, G., 11, 31
Seyle, H., 44
Shaver, P., 60, 61, 62, 158, 160, 163, 167
Shepard, P., 11, 151
Sibley, D., 91
Siegel, J., 14, 15, 58, 107
Sills, F., 116
Silverman, D., 77
Simon, A., 41, 46, 64, 88
Singer, P., 12
Skinner, E., 36, 37, 40, 51, 85
slavery, 11
Sloane, R., 129
Smith, R., 36, 40, 41, 42, 43, 84
Smith-Osborne, A., 4, 20, 40, 190, 203, 205
social class, as distal risk factor, 36
social competence
    and contact with animals, 16
    resilience and, 39, 102
social exclusion, 46, 85
social lubricant, animals' role, 13, 15
social suffering, 51
society, animals and, 7–8

262   *Index*

socioeconomic status, and risk, 36
Soren, I., 119, 195
Southam-Gerow, M., 102, 103, 192, 193
speciesism, in language, 11
Spiegal, B., 29, 30, 32, 93
sport
    emergence of therapeutic value of horses from, 19
    roots of horse-related, 18
Spradley, J., 77
Staempflif, M., 30
Stein, M., 35, 37
Steinlin, E., 6
Stewart, P., 14
stress
    effects of different stages in child development on responses to, 40
    and the immune system, 37, 84
    impact on physical health, 36
    research on resilience to, 38
Strimple, E., 16
study participants, responses to TH, *see* TH sessions
Sullivan, W., 29, 30, 175, 187
survival instinct, and mother-infant attachment, 53
Swinehart, E., 15, 112
Symington, A., 20, 33

tactile contact
    autistic spectrum and, 110, 132
    difficulty of in foster care, 108, 131
    importance of, 109–10
tactile experience, horses' facilitation, 21
Tavlou, P., 187
Taylor, A., 29, 30, 49, 175, 187
Teasdale, J. D., 59, 60, 62, 96, 157, 160, 163, 167, 168
Tedeschi, P., 13, 15, 107, 111
Teichman, M., 7, 23, 24
Terkel, J., 7, 2, 23, 24
terminology
    best practice agreements, 3
    definitions, 2–3
Tester, K., 10

Thomas, G., 31
Thompson, J., 31
Thompson, K., 105, 112
Thompson, R., 102
Thorne, B., 51, 56, 116, 152
TH sessions
    being in nature makes study participants feel calm, 170–1
    the calming influence of horses, 156–7
    Cinderella builds relationship with Leo, 113–14
    Cinderella imagines Duchess has been touched by 'unicorn dust', 182
    Cinderella learns about the importance of diet and exercise, 185–6
    Cinderella learns mindfulness with Duchess, 164–7
    Cinderella relates to Duchess, 143
    Cinderella's attachment to Leo, 115–17
    Emma envies the horses' freedom, 181–2
    Emma identifies with Jason's 'naughty' side, 143–5, 151
    Emma learns about the importance of diet and exercise, 185
    Emma modifies her behaviour, 100–1
    Emma understands the importance of the natural environment for the horses, 172–3
    experiential nature, 76–7
    field maintenance quiz, 187
    Freya experiences 'invisible' riding with Ruby, 158–61
    Freya experiences join-up with Hector, 93–5
    introjections, 152
    Jason is liked because he is a 'challenge', 183
    Kelly builds relationship with Ruby, 127–8
    Kelly builds relationships through unconditional positive regard, 127–9

Kelly discusses potential positive opportunities, 99
Kelly experiences wildlife in the natural environment, 179
Kelly feels Ruby understands her, 121
Kelly is more relaxed after TH sessions, 137
Kelly opens up, 134–6
Kelly tries 'invisible riding' on Ruby, 89–91
Lucy, Emma and Wayne choose animal-based placements, 93
Lucy and Freya show interest in nature, 176–7
Lucy feels accepted by the horses, 130
Lucy feels confident when riding, 183
Lucy feels like Queen of the world, 91
Lucy feels safe with Duchess, 123–4
Lucy gains confidence through understanding horse behaviour, 95–6
Lucy helps Timmy to overcome fear, 114
Lucy speaks of the importance of listening to your instincts, 180
Lucy's feelings of exclusion, 92
Minimax builds relationship and attachment with Ruby, 108
Minimax builds relationship with Jason, 145–7
Minimax displays acts of helpfulness, 87–9
Minimax expresses affection for Ruby, 109
Minimax feels really strong, 91
Minimax feels Ruby understands him, 120–1
Minimax finds comfort with Ruby, 130–2
Minimax gets to know Ruby, 86–7
Minimax identifies and connects with Ruby, 118–19, 142
Minimax is calm after TH sessions, 181

Minimax learns transferable skills from controlling Ruby's behaviour, 96–7
Minimax 'loves being here!', 175
Minimax modifies his behaviour, 141–2
Minimax relaxes with Ruby, 173–4
Minimax's risk-taking behaviour, 187
Minnie helps Lucy to build relationships with others, 177–8
opening up of positive opportunities for Emma, Lucy, Kelly and Wayne, 98
overview, 72–3
trauma of Kelly's last session, 139
Wayne completes worksheet, 184–5
Wayne expresses affection for Ruby, 109
Wayne finds being in nature with the horses relaxing, 174–5
Wayne identifies with Billy, 150–1
Wayne is calm and confident with Ruby, 162–3
Wayne is more relaxed after TH sessions, 137
Wayne overcomes fear in a group outing, 97
Wayne relates to Missy, 147–8
Wayne's experience of 'holding' with Ruby, 110
working outside with the horses, 186–7
see also EAT/L and TH
therapeutic gardening project
use of metaphor and analogy, 138
see also therapeutic horticulture
therapeutic horsemanship questionnaire, 216–17
therapeutic horticulture
comparison with TH, 138–41
perceived benefits, 30, 179
therapeutic interventions
usefulness in providing resilience and protective factors, 84
value of pets as, 13
therapeutic relationship
adult's views, 136–8

264 Index

therapeutic relationship – *continued*
applicability of the Winnecottian holding space, 126, 136–7
calmness, safety and trust, 136–8
client–therapist parallels with horses, 125
community interventions and, 50
horses and a safe space, 132–6
horses and the, 57–8
horses and unconditional positive regard, 127–32
horses' effectiveness, 126
identifying with the horses, 142–3
indicators of a successful, 126–7
non-directive approach, 149
quality of the, 125
role of metaphor and analogy, 138–42
Tight, M., 64, 77
torture victims, therapeutic gardening project, 138
Tottle, S., 158
traditional interventions, community-based, 49
transference, 149
traumatic events
emotional and behavioural symptoms attributed to exposure, 37
impact of early exposure to, 83–4
neurobiological perspective, 44
Travers, M., 63
Tregaskis, C., 197
Trienbenbacher, S., 15
Trotter, K., 7, 20, 33, 127, 143, 149
true self, 54–6, 116
trust
and attachment, 120
and the trauma victim's capacity for recovery, 134
Turner, D., 15
Turner, W., 13, 106
Twigg, J., 199

unconditional positive regard, 50, 126–9
Ungar, M., 32, 35, 39, 48, 49, 50, 64, 84, 85, 90, 98, 200

vaulting, 21, 109, 147
Vidrine, M., 7, 20, 21, 22, 24, 33, 50, 52, 55, 57, 102, 126, 128, 136, 141, 143, 147, 194
vitamin D deficiency, 29
Voight, A., 48, 93, 200

Waal de, F., 105
Wagner, E., 58, 59
Walker, E., 5, 8, 16, 18, 203
Walker, S., 200
Walkerdine, V., 62
Wals, A., 169, 174, 176, 177, 189, 193, 194
Walsh, D., 51
Walsh, F., 13, 15, 55, 107, 111, 137, 171, 191
Ward Thompson, C., 187
Waugh, M., 59, 155
Wayne, 70, 75, 82, 93, 97–8, 100, 106, 109–11, 137, 147–8, 150–1, 162–4, 168, 174–5, 184–5, 220
weather, influence on horses, 32
Webb, S., 63
Weinstein, M., 63, 75
Wells, D., 13, 15, 107
Wells, N., 29, 31, 32, 49, 187
Werner, E., 36, 40, 41, 42, 43, 47, 83, 84, 85, 89
Whitaker, R., 29, 32
Widdicombe, S., 67
Wilbert, C., 8, 9, 10
wild garlic, 176–7
wildness, horses as representative of, 18
Wilkie, R., 8, 10
Wilkins, G., 16, 28, 177
Williams, J., 20, 24, 27, 60, 157, 163, 167, 168
Williams, M., 63
Wilson, C., 14, 15, 107
Wilson, E. O., 3, 11, 16, 27, 28, 177
Winnecottian holding space, applicability to the therapeutic relationship, 126, 136–7
Winnicott, D., 12, 49, 51, 54, 55, 56, 111, 116, 126, 136, 152
Winter, M., 124
Wintrode, A., 183

Wodarski, J., 48, 49, 64, 200
Wolkow, K., 39, 40, 48, 49
women's rights, 11
World Health Organisation, 206
Worsham, N., 15, 112

Xenophon, 17

Yasikoff, N., 6
yoga, 58–9, 157–8
Yorke, J., 20, 55, 56, 57, 69, 105, 106, 108, 109, 126, 127, 128, 134, 193

York Retreat, 14
Young, R., 6
young people's development, impact of alienation from nature on, 28

Zeanah, C., 52
Zeltzer, L., 36, 38
Zimmer-Gembeck, M., 36, 37, 40, 51, 85
Zylowska, L., 59, 61, 157, 166, 168

CPSIA information can be obtained
at www.ICGtesting.com
Printed in the USA
LVHW081322050323
740957LV00007B/969